NEW NONLINEAR OPTICAL MATERIALS: THEORETICAL RESEARCH

NEW NONLINEAR OPTICAL MATERIALS: THEORETICAL RESEARCH

J.P. HUANG AND K.W. YU

Nova Science Publishers, Inc.
New York

NOTICE TO THE READER

The Publisher has taken reasonable care in the preparation of this book, but makes no expressed or implied warranty of any kind and assumes no responsibility for any errors or omissions. No liability is assumed for incidental or consequential damages in connection with or arising out of information contained in this book. The Publisher shall not be liable for any special, consequential, or exemplary damages resulting, in whole or in part, from the readers' use of, or reliance upon, this material.

Independent verification should be sought for any data, advice or recommendations contained in this book. In addition, no responsibility is assumed by the publisher for any injury and/or damage to persons or property arising from any methods, products, instructions, ideas or otherwise contained in this publication.

This publication is designed to provide accurate and authoritative information with regard to the subject matter cover herein. It is sold with the clear understanding that the Publisher is not engaged in rendering legal or any other professional services. If legal, medical or any other expert assistance is required, the services of a competent person should be sought. FROM A DECLARATION OF PARTICIPANTS JOINTLY ADOPTED BY A COMMITTEE OF THE AMERICAN BAR ASSOCIATION AND A COMMITTEE OF PUBLISHERS.

Library of Congress Cataloging-in-Publication Data
Huang, Ji-Ping.
 New nonlinear optical materials : theoretical research / Ji-Ping Huang.
  p. cm.
 Includes index.
 ISBN 13: 978-1-60021-402-8
 ISBN 10: 1-60021-402-9
1. Nonlinear optics–Materials. 2. Optical materials. I. Title.
QC446.2.H83   2006
621.36'94–dc22   2006025059

Published by Nova Science Publishers, Inc. ✤ *New York*

Contents

Preface

Optical materials with large values of nonlinear susceptibilities and fast responses are in great need in industrial applications, such as nonlinear optical switching devices for use in photonics and real-time coherent optical signal processors, optical limiters, and so on. In general, many applications of nonlinear optics that have been demonstrated under controlled laboratory conditions could become practical for technological uses if such materials were available. It is usually believed that an effective enhanced nonlinear optical response can appear in a composite material in which at least one component should possess an inherent nonlinear optical response. Thus, the common way to develop new nonlinear optical materials is to seek materials in which the components possess an inherently large nonlinear optical response. In contrast, we have theoretically exploited some new nonlinear optical materials, e.g., colloidal nanocrystals with strong lattice effects, metallic films with inhomogeneous microstructures adjusted by ion doping or temperature gradient, composites of graded (and/or shape-anisotropic) nanoparticles, etc. The proposed materials are difficult or impossible to achieve with conventional, naturally occurring materials or random composites widely discussed in the literature. Recently, our group has made substantial progress along this line. In this regard, we are in a position to write a comprehensive book, which is an original and first-handed review of the state-of-the-art development of the field, in order to show the reader the up-to-date information.

So far, there are mainly four competitive books that are respectively:

(1) Y. R. Shen, The Principles of Nonlinear Optics (John Wiley and Sons, New York, 1984).

(2) P. N. Butcher and D. Cotter, The Elements of Nonlinear Optics (Cambridge University Press, New York, 1990).

(3) R. W. Boyd, Nonlinear Optics (Academic Press, New York, 1992).

(4) V. M. Shalaev, Nonlinear optics of random media: Fractal composites and metal-dielectric films (Springer, Berlin, 2000).

We believe that these books do not affect the significance of our book. Details: (1) offers a comprehensive treatment of nonlinear optics emphasizing physical concepts and the relationship between theory and experiment, and systematically describes a number of sub-topics in the field. It contains almost every fundamental aspect in the field of nonlinear optics. (2) introduces the principles of nonlinear optics and quantum mechanical apparatus. Some of the topics of (3) include the fundamentals and applications of optical systems based on the nonlinear interaction of light with matter. Other topics of (3) include mechanisms of optical nonlinearity, second-harmonic and sum- and difference-frequency generation, photonics and optical logic, optical self-action effects including self-focusing and optical soliton formation, optical phase conjugation, stimulated Brillouin and stimulated Raman scattering, and selection criteria of nonlinear optical materials. In a word, the focus of (1)-(3) is not on developing new nonlinear optical materials. All of them mainly focus on the theories that describe the mechanisms or principles of nonlinear optics. This differs from our book that systematically describes the theoretical progress in developing new nonlinear optical materials. (4)'s aim is on exploiting random media such as metal-dielectric composites or films, which is substantially different from the media of our interest, too. In a word, the aim of our book is to develop/establish theories to design new nonlinear optical materials, which seems to be quite different from (1)-(4). In fact, besides the competitive books listed above, there are also some relevant conference proceedings. However, they may provide a broader perspective, but usually lack continuity. Apparently, this book is not only advanced, but also continuous due to its self-consistence.

The subject of our book is to design new nonlinear optical materials by developing or establishing various theoretical techniques, such as first-principles approaches, the Ewald method, effective medium theories, the differential effective dipole theory, and so on. The focus of the book is not only on theoretical side alone, but also on potential application of the theoretical designs. Thus, this book is expected to be of value for both experimentalists (in academy or industry) and theorists, and receive a broad interest in the engineering, physics and optics communities. Since the written language in the book is collegial, it is also readable and understandable for beginners and/or graduate students.

For completing the book we have profited from valuable and stimulating collaborations and discussions with Prof. G. Q. Gu, Prof. L. Gao, Dr. M. Karttunen, Prof. K. Yakubo, Prof. T. Nakayama, Prof. C. Holm, Prof. P. M. Hui, Mr. J. J. Xiao, and Ms. L. Dong. J.P.H. would also like to express his gratitude to Ms. C. Z. Fan, Mr. G. Wang, Mr. W. J. Tian, and Ms. Y. J. Zhao for their helpful assistance. We acknowledge the financial support by the Research Grants Council of the Hong Kong SAR Government, by the Alexander von Humboldt Foundation in Germany, by the German Research Foundation under Grant No. HO 1108/8-4, by the Grant-in-Aid for Scientific Research organized by Japan Society for the Promotion of Science, by the Shanghai Education Committee and the Shanghai Education Development Foundation ("Shu Guang" project), by the Scientific Research Foundation for the Returned Overseas Chinese Scholars, State Education Ministry, China, and by the Department of Physics, Fudan University, China,

J. P. Huang
Professor
Department of Physics, Fudan University, Shanghai 200433, P. R. China.
Email: jphuang@fudan.edu.cn

K. W. Yu
Professor
Department of Physics and Institute of Theoretical Physics, The Chinese University of Hong Kong, Shatin, New Territories, Hong Kong.
Email: kwyu@phy.cuhk.edu.hk

January, 2006

Chapter 1

Introduction

Nonlinear optics is the basis of all the fledgling photonics technologies, where light works, or even replaces, electrons in applications traditionally carried out by microelectronics. Actually, the field of nonlinear optics traces its beginning to 1961, when a ruby laser was first used to generate the second-harmonic radiation inside a crystal [1]. The realization of all-optical switching, modulating and computing devices is an important goal in modern optical technology. Nonlinear optical materials with large third-order nonlinear susceptibilities are indispensable for such devices, because the magnitude of this quantity dominates the device performance. In general, many applications of nonlinear optics that have been demonstrated under controlled laboratory conditions could become practical for technological uses if such materials were available. Thus, finding nonlinear optical materials with a large third-order nonlinear optical susceptibility is up to now a challenge [2, 3, 4, 5, 6, 7]. The most common way to achieve this is to use composite materials in which the constituent components possess large intrinsic nonlinear responses. Noble metal (typically gold, silver and copper) is often chosen as an ingredient due to their extremely large and fast nonlinear optical response. Many different microstructures have been exploited in an attempt to access the intrinsic optical nonlinearity of metals, for example, the random metallodielectric composites [8, 9, 10], fractal films [9, 10], alternative bilayers [4, 5, 11], etc. They basically rely on the enhanced local fields in space or on the effectively lengthened scale of the interactions between the matter and the incident light field. The interaction between an incident light (a kind of electromagnetic wave) and a material can be described by the Maxwell equations that contain material's relative complex dielectric constant. If the particles in a material are conductive and connected, there is a flow of conducting electrons, Ohmic current, through the system. There is also a dipolar response, which arises in the Maxwell equations through the displacement current. For a Drude metal, the Ohmic current dominates in the low-frequency region. In contrast, the displacement current (dipolar response) dominates in the high-frequency region. In this region, there is a non-compensated surface charge on particles resulting in their polarization or dipolar response, and thus changing the field acting on the particles.

Moreover, there is also a great demand for particular optical materials in devices applications, which would benefit from additional tunability of the optical properties. Recently, we studied graded composites, which provided an extra degree of freedom for controlling the optical properties of these materials [12, 13, 14, 15, 16]. In fact, there exist in Nature abundant graded materials, such as biological cells [17] and liquid crystal droplets [18]. After all, many artificially-graded-index optical metamaterials and elements have been fabricated nowadays [19]. For the research on nonlinear optical responses, the introduction of a controllable element (e.g., external magnetic field or gradation) should be expected to open a fascinating field of new phenomena.

The nonlinear optical properties of nanoscale composite materials are often quite different from the properties of the constituent materials from which the composite is constructed. Composite materials can have larger nonlinear susceptibilities at zero and finite frequencies than those of ordinary bulk materials. The formation of composite materials thus constitutes a means for engineering new materials with desired nonlinear optical properties [20]. The authors [21] studied nonlinear optical properties of fractal aggregates and showed that the aggregation of initially isolated particles into fractal clusters results in a huge enhancement of the nonlinear response within the spectral range of collective dipolar resonances like surface plasmon resonances. Basically, the response of a nonlinear composite can be tuned by controlling the volume fraction and morphology of constitutes.

Typically, the nanoparticle size ranges from tens to hundreds of nanometers. The particles are embedded in a host material and can be aggregated into chains (or clusters) under different conditions. When a collection of objects (e.g., nanoparticles or nanoparticle chains) whose size and spacing are much smaller than the wavelength of an incident light, the light passing through the structure cannot tell the difference, and hence the inhomogeneous structure can be seen as a homogeneous one [22]. In this case, the quasi-static approximation can be used to describe the optical response of an individual particle or a nanoparticle chain. In other words, to investigate the nonlinear optical responses of the proposed material, we are allowed to average over inhomogeneous nanoparticles or nanoparticle chains, conceptually replacing the inhomogeneous objects by a homogeneous material.

There are a number of optical processes, e.g., four-wave mixing, second or third harmonic generation, etc. It is important to keep in mind that, because of the symmetry requirement, all even order susceptibilities vanish when the material is centrosymmetric. In comparison, odd order optical susceptibilities are possible in all systems.

In this book, we shall present an original, and first-handed review of the state-of-the-art development on the design of new nonlinear optical materials, with an emphasis on understanding the physical processes of the composite effects on the enhancement of optical nonlinearity of the materials.

The book is organized as follows. First, Chapter 2 presents three basic theories that will be used in other chapters. Then, in Chapter 3, we theoretically design a class of nonlinear optical materials, colloidal nanocrystals with graded particles or a graded-index host. We design inhomogeneous metallic films as new nonlinear optical materials in Chapter 4. In Chapter 5, graded composites are further investigated to design new nonlinear optical materials based on them. Chapter 6 presents a kind of magneto-controlled ferrofluid-based nonlinear optical materials, whose microstructure can be tuned by an external magnetic field. In Chapter 7, we shall also theoretically show that electrorheological nanofluids and ferrofluids themselves can serve as new nonlinear optical materials when they are partially subjected to a Gaussian laser beam (light-striction). A survey on the design of nonlinear optical materials by other materials will be done in Chapter 8. This follows by Chapter 9 where a summary is presented.

Chapter 2

Basic Theories

Composites often contain a macroscopic scale of inhomogeneity. In such a material, there are small, yet much larger than atomic, regions where macroscopic homogeneity prevails and where the foregoing macroscopic parameters suffices to characterize the physics, but different regions may have quite different values for those parameters. If we are interested in the physical properties at scales that are much larger than those regions and at which the material appears to be homogeneous, then the macroscopic behavior can again be characterized by bulk effective values, e.g. effective dielectric constant ε_e. In what follows, we shall review three typical theories for calculating ε_e, namely, the Maxwell-Garnett theory (Section 1.), the Bruggeman theory (Section 2.) as well as the Bergman-Milton spectral representation theory (Section 3.).

1.   The Maxwell-Garnett Theory

The Maxwell-Garnett theory [23, 24] (or Maxwell-Garnett approximation) is also known as the Clausius-Mossotti theory. Regarding how to derive the equation for the Maxwell-Garnett theory, there are several approaches. Here we would like to start from the view of effective local electric field.

Let us discuss a two-component composite where many particles of the dielectric constant ε_1 and the volume fraction p are randomly embedded in a host medium of ε_2, in the

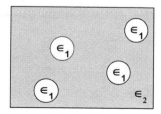

Figure 2.1. Schematic graph showing an asymmetrical microstructure for the Maxwell-Garnett theory in which component 1 with dielectric constant ε_1 is embedded in component 2 with ε_2.

presence of an external electric field E_0 along the z-axis. Then, we denote the local electric field inside the particle by E_1, and that inside the host medium by E_2. Solving a fundamental equation in electrodynamics yields

$$E_1 = \frac{3\varepsilon_2}{\varepsilon_1 + 2\varepsilon_2} E_2. \tag{2.1}$$

It is known that the average electric field $\langle E \rangle (\equiv p\langle E_1 \rangle + (1-p)\langle E_2 \rangle)$ inside the composite should be equal to the external electric field, namely, $\langle E \rangle = E_0$. On the other hand, the effective dielectric constant ε_e may be given by the ratio of the average displacement $\langle D \rangle$ to the average electric field $\langle E \rangle$ inside the composite , namely,

$$\varepsilon_e = \frac{\langle D \rangle}{\langle E \rangle} = \frac{\langle D \rangle}{E_0}. \tag{2.2}$$

Hence we obtain

$$\varepsilon_e = \frac{p\varepsilon_1\langle E_1 \rangle + (1-p)\varepsilon_2\langle E_2 \rangle}{p\langle E_1 \rangle + (1-p)\langle E_2 \rangle}. \tag{2.3}$$

To this end, we obtain the expression for the Maxwell-Garnett theory as

$$\varepsilon_e = \varepsilon_2 \frac{\varepsilon_1(1+2p) + 2\varepsilon_2(1-p)}{\varepsilon_1(1-p) + \varepsilon_2(2+p)}. \tag{2.4}$$

We re-express Eq.(2.4) in a commonly-used form as

$$\frac{\varepsilon_e - \varepsilon_2}{\varepsilon_e + 2\varepsilon_2} = p\frac{\varepsilon_1 - \varepsilon_2}{\varepsilon_1 + 2\varepsilon_2}. \tag{2.5}$$

Obviously, the Maxwell-Garnett theory is an asymmetrical theory (see Fig. 2.1), namely the physical property of the composite can be changed if one exchanges the notations 1 and 2. Finally, for the Maxwell-Garnett theory the extension to multi-component composite is straightforward as can be easily done on the same footing.

2. The Bruggeman Theory

Another approach to calculating ε_e for a two-component composite similar to the above was introduced by Bruggeman [25]. Thus, this approach is called the Bruggeman theory (also called the effective medium theory or Bruggeman approximation). Its predictions are usually sensible and physically offer a means of quick insight into some problems that are difficult to attack by other approaches.

We would derive the expression for the Bruggeman theory by considering the fact that the effective dipole factor $\langle b \rangle$ of the composite should be zero, namely

$$\langle b \rangle = 0. \tag{2.6}$$

On the other hand, we have

$$\langle b \rangle = pb_1 + (1-p)b_2, \tag{2.7}$$

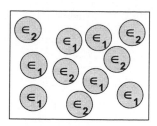

Figure 2.2. Schematic graph showing a symmetrical microstructure for the Bruggeman theory in which two components are mixed.

where b_1 and b_2 are respectively the dipole factors of the particle and the host medium in the effective medium, and they are given by

$$b_1 = \frac{\varepsilon_1 - \varepsilon_e}{\varepsilon_1 + 2\varepsilon_e}, \text{ and } b_2 = \frac{\varepsilon_2 - \varepsilon_e}{\varepsilon_2 + 2\varepsilon_e}. \tag{2.8}$$

In view of Eqs.(2.6) - (2.8), we have directly the expression for the Bruggeman theory

$$p\frac{\varepsilon_1 - \varepsilon_e}{\varepsilon_1 + 2\varepsilon_e} + (1-p)\frac{\varepsilon_2 - \varepsilon_e}{\varepsilon_2 + 2\varepsilon_e} = 0. \tag{2.9}$$

Obviously, the Bruggeman theory is a symmetrical theory (see Fig. 2.2), namely the physical property of the composite can keep unchanged if one exchanges the notations 1 and 2. In addition, for the Bruggeman theory the extension to multi-component composite is also straightforward which can be readily done on the same footing as above.

3. The Bergman-Milton Spectral Representation Theory

For the sake of convenience, we let $\mathbf{E}_0 = -\hat{\mathbf{e}}_z$. Below we briefly review the Bergman-Milton spectral representation theory for the effective dielectric constant of a two-component composite.

The problem is initiated by solving the differential equation [26]

$$\nabla \cdot [(1 - \frac{1}{s}\eta(\mathbf{r}))\nabla\phi(\mathbf{r})] = 0, \tag{2.10}$$

where $s = \varepsilon_2/(\varepsilon_2 - \varepsilon_1)$ denotes the relevant material parameter and $\eta(\mathbf{r})$ is the characteristic function of the composite, having value 1 for r in the embedding medium and 0 otherwise. The electric potential $\phi(\mathbf{r})$ can be solved formally

$$\phi(\mathbf{r}) = z + \frac{1}{s}\int d\mathbf{r}'\eta(\mathbf{r}')\nabla'G_0(\mathbf{r}-\mathbf{r}')\cdot\nabla'\phi(\mathbf{r}'), \tag{2.11}$$

where $G_0(\mathbf{r}-\mathbf{r}') = |\mathbf{r}-\mathbf{r}'|/4\pi$ is the free space Green's function. By denoting an integral-differential operator

$$\Gamma = \int d\mathbf{r}'\eta(\mathbf{r}')\nabla'G_0(\mathbf{r}-\mathbf{r}')\cdot\nabla', \tag{2.12}$$

and the corresponding inner product

$$< \phi|\Phi >= \int d\mathbf{r}\eta(\mathbf{r})\nabla\phi^* \cdot \nabla\Phi. \tag{2.13}$$

It is easy to show that Γ is a Hermitian operator. Let s_n and $\Phi_n(\mathbf{r})$ be the $n-$th eigenvalue and eigenfunction of the Γ operator, respectively, then we obtain the effective dielectric constant ε_e in the Bergman-Milton spectral representation

$$\begin{aligned}
\varepsilon_e &= -\frac{1}{V}\int dV\varepsilon(\mathbf{r})E_z \\
&= \frac{1}{V}\int dV\varepsilon_2\left[1-\frac{1}{s}\eta(\mathbf{r})\right]\frac{\partial\Phi}{\partial z} \\
&= \varepsilon_2\left(1-\frac{1}{V}\sum_n\frac{|<\Phi_n|z>|^2}{s-s_n}\right) \\
&= \varepsilon_2\left(1-\sum_n\frac{F_n}{s-s_n}\right).
\end{aligned} \tag{2.14}$$

The parameters s_n and F_n satisfy simple properties that $0 \leq s_n \leq 1$ and $\sum F_n = p$ [26]. Eq. (2.14) is just the effective dielectric constant of a two-component system in the Bergman-Milton spectral representation. Moreover, after introducing $F(s)$ as a function of s as

$$F(s) \equiv \sum_n\frac{F_n}{s-s_n}, \tag{2.15}$$

we may readily obtain the spectral structure of the composite. In doing so, we may further represent $F(s)$ as

$$F(s) = \int_0^1 dx\frac{\mu(x)}{s-x}, \tag{2.16}$$

where the spectral function $\mu(x)$ is a crucial parameter which contains the information about the spectral structure, and is thus given by

$$\mu(x) = -\frac{1}{\pi}\mathrm{Im}F(x+i0^+). \tag{2.17}$$

Obviously, we may observe that the Bergman-Milton spectral representation is a rigorous mathematical formalism for the effective dielectric constant of a two-phase composite material [26]. It offers the advantage of the separation of materials parameters (namely the dielectric constant or conductivity) from the particle structure information (see Eq. (2.14)), thus simplifying the study.

For a better understanding of the Bergman-Milton spectral representation, below we would like to represent the above-mentioned Maxwell-Garnett and Bruggeman theories in the Bergman-Milton spectral representation. As a result, for the Maxwell-Garnett theory we respectively have the $F(s)$ function and the spectral function as

$$F(s) = \frac{p}{s-(1-p)/3} \text{ and } \mu(x) = p\delta[x-(1-p)/3].$$

On the other hand, for the Bruggeman theory we have the $F(s)$ function as

$$F(s) = \frac{1}{4s}\left(-1 + 3p + 3s - 3\sqrt{(s-x_1)(s-x_2)}\right),$$

where x_1 and x_2 are given by solving

$$(1-3p)^2 - 6(1+p)x + 9x^2 = 0,$$

hence,

$$x_1 = \frac{1}{3}\left(1 + p - 2\sqrt{2p(1-p)}\right) \text{ and } x_2 = \frac{1}{3}\left(1 + p + 2\sqrt{2p(1-p)}\right).$$

In this case, the spectral function should be

$$\mu(x) = \frac{3p-1}{2}\theta(3p-1) + \frac{3}{4\pi x}\sqrt{(x-x_1)(x_2-x)}$$

as $x_1 < x < x_2$, and

$$\mu(x) = \frac{3p-1}{2}\theta(3p-1)$$

otherwise.

We would also like to mention that the extension of the Bergman-Milton spectral representation to the three-component composite can be made by taking into account various approaches [27].

In Appendix A, we offer an easy-to-understand illustration of the Bergman-Milton spectral representation by using the capacitance of simple geometry [28].

Chapter 3

Colloidal Nanocrystals

Colloidal crystalline is extensively studied in nanomaterials engineering and its potential applications range from nanophotonics to chemistry and biomedicine [29]. Colloidal crystals can be prepared via templated sedimentation, methods based on capillary forces, and electric fields [30, 31, 32]. They exhibit body centered tetragonal (bct), body centered cubic (bcc) and face-centered cubic (fcc) structures, depending on the lattice constants and hence the volume fraction of colloidal particles. These structures can be investigated by using static and dynamic light scattering techniques [33, 34]. So far, colloidal-based optical sensors [35] (and photonic-band-gap materials based on inverse opaline structures [36]) have been made possible by these fabrication techniques.

This Chapter describes a class of colloidal-crystal-based nonlinear optical materials, which are made of graded metallodielectric nanoparticles (namely, a graded metallic core plus a dielectric shell), or a graded-index host.

1.   With Graded Particles

Let us start by considering a tetragonal unit cell which has a basis of two colloidal nanoparticles each of which is fixed with an induced point dipole at its center. One of the two nanoparticles is located at a corner and the other one at the body center of the cell (Fig. 3.1). Its lattice constants are denoted by $c_1(=c_2) = \ell q^{-1/2}$ and $c_3 = \ell q$ along $x(y)$ and z axes, respectively. In this case, the uniaxial anisotropic axis is directed along z axis. The degree

Table 3.1. The volume fraction p of the metallic component for bct (i.e., $q = 0.87358$ or $\alpha_{\parallel} = 1.09298$), bcc (i.e., $q = 1.0$ or $\alpha_{\parallel} = 1.0$) and fcc (i.e., $q = 2^{1/3}$ or $\alpha_{\parallel} = 1.0$) lattices at various thickness parameter t.

	$t = 1.2$	2.0	3.0
bct	p=0.40401	0.08727	0.02586
bcc	p=0.39362	0.08502	0.02519
fcc	p=0.42852	0.09256	0.02743

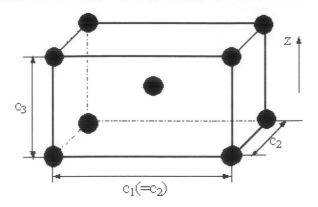

Figure 3.1. Schematic graph showing the location of two colloidal nanoparticles in a tetragonal unit cell. After Ref. [37].

of anisotropy of the periodic lattice is measured by how q deviates from unity. In particular, $q = 0.87358$, 1.0 and $2^{1/3}$ represents the bct, bcc and fcc lattice, respectively. In general, for a colloidal crystal, the individual colloidal nanoparticles should be touching. In fact, a colloidal crystal without the particles' touching can also be made if the colloidal nanoparticles are charged and stabilized by electrostatic forces. In this section we shall investigate colloidal crystals with the particles' touching.

With recent advancements in the fabrication of nanoshells [38, 39], we are allowed to use a dielectric surface layer with thickness d on a graded metallic core with radius a, in order to activate repulsive (or attractive) force between the nanoparticles. This is also a crucial requirement because otherwise multipolar interaction between the metallic cores can be important. The dielectric constant $\varepsilon_1(r)$ $(r \leq a)$ of the metallic core should be a radial function, because of a radial gradation. The dielectric constant ε_s of the surface layer can be the same as that ε_2 of the host fluid, as to be used in the following. In this regard, the surface layer contributes to the geometric constraint

$$c_1{}^2 + c_2{}^2 + c_3{}^2 = 16(a+d)^2, \tag{3.1}$$

rather than the effective optical responses. Owing to this constraint, it is found that the smallest q occurs at the bct lattice while the largest q occurs at the fcc. Meanwhile, we obtain a relation between q and the volume fraction p of the metallic component,

$$p = \frac{\pi}{24t^3}\left(\frac{q^3+2}{q}\right)^{3/2}, \tag{3.2}$$

with thickness parameter $t = (a+d)/a$. When an external electric field $\mathbf{E}_0$ is applied along x axis, the induced dipole moment $\mathbf{P}$ are perpendicular to the uniaxial anisotropic axis. Then, the local field $\mathbf{E}_L$ at the lattice point $\mathbf{R} = \mathbf{0}$ can be determined by using the Ewald-Kornfeld formulation [40],

$$E_L = P \sum_{j=1}^{2} \sum_{\vec{R} \neq \vec{0}} [-\gamma_1(R_j) + x_j^2 q^2 \gamma_2(R_j)] - \frac{4\pi P}{V_c} \sum_{\vec{G} \neq \vec{0}} \Pi(\vec{G}) \frac{G_x^2}{G^2} \exp\left(\frac{-G^2}{4\eta^2}\right) + \frac{4P\eta^3}{3\sqrt{\pi}}. \tag{3.3}$$

Here we have taken into account the influence of the local-field effect arising from all the other particles in the lattice (*lattice effect*). In Eq. (3.3), γ_1 and γ_2 are two coefficients, given by

$$\gamma_1(r) = [\text{erfc}(\eta r)/r^3] + (2\eta/\sqrt{\pi}r^2)\exp(-\eta^2 r^2), \tag{3.4}$$

$$\gamma_2(r) = [3\text{erfc}(\eta r)/r^5] + [4\eta^3/(\sqrt{\pi}r^2) + 6\eta/(\sqrt{\pi}r^4)]\exp(-\eta^2 r^2), \tag{3.5}$$

where $\text{erfc}(\eta r)$ is the complementary error function, and η an adjustable parameter making the summation converge rapidly. R and G denote respectively the lattice vector and the reciprocal lattice vector,

$$\vec{R} = \ell(q^{-1/2}l\hat{x} + q^{-1/2}m\hat{y} + qn\hat{z}), \tag{3.6}$$

$$\vec{G} = (2\pi/\ell)(q^{1/2}u\hat{x} + q^{1/2}v\hat{y} + q^{-1}w\hat{z}), \tag{3.7}$$

where $l, m, n, u, v,$ and w are integers. In addition, x_j and R_j are given by

$$x_j = l - (j-1)/2, \tag{3.8}$$

$$R_j = |\vec{R} - [(j-1)/2](a\hat{x} + a\hat{y} + c\hat{z})|, \tag{3.9}$$

and the structure factor $\Pi(\vec{G}) = 1 + \exp[i(u+v+w)/\pi]$. Now we define a local field factor in transverse field cases, $\alpha_\perp = 3V_c E_L/(4\pi P)$. It is worth noting that $\alpha_\perp$ is a function of a single variable q. Also, there is a sum rule [40, 41]

$$2\alpha_\perp + \alpha_\| = 3, \tag{3.10}$$

where $\alpha_\|$ denotes the local field factor in longitudinal field cases. Here the longitudinal (or transverse) field case corresponds to the fact that the E field of the incident light is parallel (or perpendicular) to the uniaxial anisotropic z axis. For the bct, bcc and fcc lattices, we obtain $\alpha_\perp = 0.95351, 1.0,$ and 1.0 (or alternatively $\alpha_\| = 1.09298, 1.0,$ and 1.0), respectively. If there are no special instructions, we shall use α to denote both $\alpha_\perp$ and $\alpha_\|$ in the following.

Next, for obtaining the effective dielectric constant ε_e (which also indicates both $\varepsilon_e^\perp$ and $\varepsilon_e^\|$, the effective dielectric constants in transverse and longitudinal field cases, respectively) of the colloidal crystal, we resort to the anisotropic Maxwell-Garnett formula [7] with a high degree of accuracy [42] due to the explicit determination of α,

$$\frac{\varepsilon_e - \varepsilon_2}{\alpha\varepsilon_e + (3-\alpha)\varepsilon_2} = p\frac{\bar{\varepsilon}_1 - \varepsilon_2}{\bar{\varepsilon}_1 + 2\varepsilon_2}, \tag{3.11}$$

where the equivalent dielectric constant $\bar{\varepsilon}_1 \equiv \bar{\varepsilon}_1(r=a)$ for the graded metallic core can be obtained [13, 43] by solving

$$d\bar{\varepsilon}_1(r)/dr = [\varepsilon_1(r) - \bar{\varepsilon}_1(r)][\bar{\varepsilon}_1(r) + 2\varepsilon_1(r)]/[r\varepsilon_1(r)], \tag{3.12}$$

as long as the gradation profile $\varepsilon_1(r)$ is given.

Assuming both the host fluid and dielectric surface layer to be linear for convenience, the effective third-order nonlinear susceptibility $\chi_e^{(3)}$ for the graded colloidal crystal is given by

$$\chi_e^{(3)} = p\frac{\langle|\mathbf{E}_{\text{lin}}|^2\rangle\langle\mathbf{E}_{\text{lin}}^2\rangle}{|\mathbf{E}_0|^2\mathbf{E}_0^2}\bar{\chi}_1^{(3)} \tag{3.13}$$

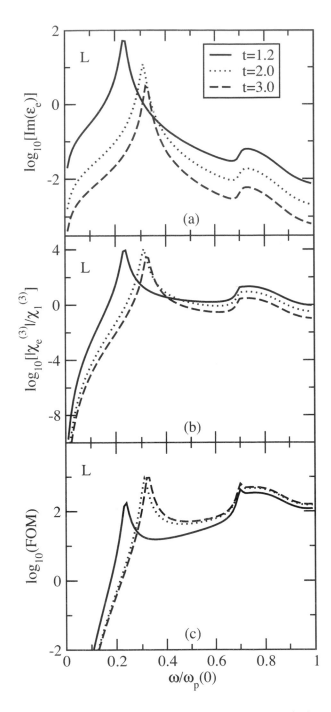

Figure 3.2. For the bct lattice, (a) the linear optical absorption $\mathrm{Im}\,(\varepsilon_e)$, (b) the enhancement of the third-order optical nonlinearity $|\chi_e^{(3)}|/\chi_1^{(3)}$, and (c) the FOM $\equiv |\chi_e^{(3)}|/[\chi_1^{(3)}\mathrm{Im}(\varepsilon_e)]$ versus the normalized incident angular frequency $\omega/\omega_p(0)$ for longitudinal field cases (L), for different t. Parameters: $\varepsilon_2 = (3/2)^2$, $C_\omega = 0.3$, and $\gamma = 0.02\omega_p(0)$. After Ref. [37].

where the equivalent third-order nonlinear susceptibility $\bar{\chi}_1^{(3)} \equiv \bar{\chi}_1^{(3)}(r = a)$ for the graded metallic core can be obtained [13] by solving

$$
\begin{aligned}
d\bar{\chi}_1^{(3)}(r)/dr &= \bar{\chi}_1^{(3)}(r)\{(3d\bar{\varepsilon}_1(r)/dr)/(2\varepsilon_2 + \bar{\varepsilon}_1(r)) + [(d\bar{\varepsilon}_1(r)/dr)/(2\varepsilon_2 + \bar{\varepsilon}_1(r))]^*\} \\
&\quad + \bar{\chi}_1^{(3)}(r)(6Y + 2Y^* - 3)/r + 3\chi_1^{(3)}(r)/(5r) \\
&\quad |(\bar{\varepsilon}_1(r) + 2\varepsilon_1(r))/(3\varepsilon_1(r))|^2 [(\bar{\varepsilon}_1(r) + 2\varepsilon_1(r))/(3\varepsilon_1(r))]^2 \\
&\quad (5 + 18X^2 + 18|X|^2 + 4X^3 + 12X|X|^2 + 24|X|^2X^2),
\end{aligned}
\tag{3.14}
$$

with

$$
X = \frac{\bar{\varepsilon}_1(r) - \varepsilon_1(r)}{\bar{\varepsilon}_1(r) + 2\varepsilon_1(r)},
\tag{3.15}
$$

$$
Y = \frac{[\varepsilon_1(r) - \varepsilon_2][\bar{\varepsilon}_1(r) - \varepsilon_1(r)]}{\varepsilon_1(r)[\bar{\varepsilon}_1(r) + 2\varepsilon_2]},
\tag{3.16}
$$

as long as the gradation profile $\chi_1^{(3)}(r)$ is also given. Here the intrinsic weak third-order nonlinear susceptibility $\chi_1^{(3)}(r)$ satisfies the local constitutive relation between the displacement $\mathbf{D}_1(r)$ and the electric field $\mathbf{E}_1(r)$ inside the graded metallic core,

$$
\mathbf{D}_1(r) = \varepsilon_1(r)\mathbf{E}_1(r) + \chi_1^{(3)}(r)|\mathbf{E}_1(r)|^2\mathbf{E}_1(r).
\tag{3.17}
$$

In Eq. (3.13), $\langle \cdots \rangle$ denotes the volume average over the metallic region, and $\mathbf{E}_{\text{lin}}$ the equivalent linear local electric field in the graded metallic core with the same gradation profile but with a vanishing nonlinear response at the frequency concerned. Both $\langle |\mathbf{E}_{\text{lin}}|^2 \rangle$ and $\langle \mathbf{E}_{\text{lin}}^2 \rangle$ can be expressed in the spectral representation as [44],

$$
\langle |\mathbf{E}_{\text{lin}}|^2 \rangle = (\mathbf{E}_0^2/p) \int dx' |s|^2 \mu(x')/|s - x'|^2,
\tag{3.18}
$$

$$
\langle \mathbf{F}_{\text{lin}}^2 \rangle = (\mathbf{E}_0^2/p) \int dx' s^2 \mu(x')/(s - x')^2,
\tag{3.19}
$$

with material parameter $s = \varepsilon_2/(\varepsilon_2 - \bar{\varepsilon}_1)$, where $\mu(x')$ is the spectral density. Since the ε_e in Eq. (3.11) can be expressed as

$$
\varepsilon_e = \varepsilon_2 \left[1 - \frac{p}{s - (1 - p\alpha)/3}\right] \equiv \varepsilon_2[1 - F(s)],
\tag{3.20}
$$

according to

$$
\mu(x') = -\frac{1}{\pi}\text{Im}[F(x' + i0^+)]
\tag{3.21}
$$

we obtain

$$
\mu(x') = p\delta\left(x' - \frac{1 - p\alpha}{3}\right).
\tag{3.22}
$$

Here $\text{Im}[\cdots]$ denotes the imaginary part of $\cdots$.

The point of achieving the resonant plasma band shown in Figs. 3.2-3.3 is that one needs a sufficiently large gradient rather than a crucially particular form of the dielectric function or gradation profiles. To show the effects of gradation, we have adopted the Drude form

$$
\varepsilon_1(r) = 1 - \frac{\omega_p^2(r)}{\omega(\omega + i\gamma)}
\tag{3.23}
$$

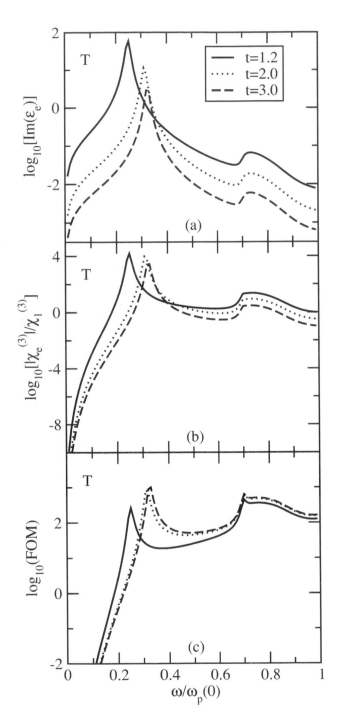

Figure 3.3. Same as Fig. 3.2, but for transverse field cases (T).

with a model plasma-frequency gradation profile

$$\omega_p(r) = \omega_p(0) \left(1 - C_\omega \frac{r}{a} \right). \tag{3.24}$$

One possible way to achieve such gradation is to fabricate a graded metallic core by using different noble metals as different layers inside the core. For focusing on the nonlinearity enhancement, we consider a model system where $\chi_1^{(3)}(r) = \chi_1^{(3)}$ is a real and positive frequency-independent constant and does not have a gradation profile. In this case, the equivalent nonlinear susceptibility $\bar{\chi}_1^{(3)}(r)$ should still depend on r because of the radial function $\varepsilon_1(r)$ [13].

For a given thickness of shell, when q varies from bct to bcc, to fcc lattices, the volume fraction p first decreases from bct, reaches a minimum at bcc, then increases again towards fcc (Table 3.1). At the same time, the longitudinal local field factor $\alpha_\parallel$ decreases from 1.09 at bct lattices almost monotonically to 1 at bcc and fcc lattices. Thus, for the bct case, the large p and large $\alpha_\parallel$ should give rise to a large red shift (namely, the plasma resonant peak and band are caused to be located at lower frequencies) from the single particle case with $p \approx 0$ where the lattice effect disappears and only the gradation effect exists. For bcc lattices, the red shift should be the smallest due to small p and $\alpha_\parallel = 1$, and hence for fcc lattices, the red shift should lie between those of bct and bct lattices. This is because α does not change much while p changes significantly (Table 3.1). From Fig. 3.2, it is evident that for a given lattice, the effective linear and nonlinear responses depend strongly on the thickness parameter t. Both the redshift and strength of the plasma resonant peak or band is largest at smallest t [Fig. 3.2(a)-(b)]. This is a combination of the local field effect and the volume fraction effect in the colloidal crystal. However, for a given thickness parameter t, the dependence of these responses on the crystal structure is not prominent (no figures shown here). In fact, the plasma resonant band in Fig. 3.2 is caused to appear by the gradation, as discussed in Ref. [14] in which the results for the case of various C_ω (and hence various degree of gradation) have also been reported. Similar results are displayed in Fig. 3.3 where the transverse field cases are investigated. In comparison with the longitudinal field cases (Fig. 3.2), the framework of the responses in transverse field cases (Fig. 3.3) is slightly blue-shifted (i.e., located at higher frequency). Owing to the volume fraction effect, this behavior is more evident for the smallest t, but small enough to be neglected at the largest t. The difference between the results predicted in Figs. 3.2 and 3.3 is generally small because $\alpha_\parallel = 1.09298$ is so close to $\alpha_\perp = 0.95351$. However, the small shift can actually be detected in experiments.

We believe dielectrophoresis can offer a convenient way of preparing a colloidal crystal [45]. It is not unusual that one fabricates graded colloidal crystals by dielectrophoresis. In this case, the particles in a medium may have different dielectric properties but they must be of the same size (so as to form colloidal crystal). In a nonuniform applied field, the different particles experiences different dielectrophoretic forces according to their strength of polarization. In this way, one really fabricates graded colloidal crystals whose dielectric properties may vary along one direction. We have considered uniform colloidal crystals of graded metallic particles. Regarding the fabrication of graded metallic spheres, a practical choice in experiments might possibly be to fabricate multilayered particles with or without dielectric anisotropy [46, 47]. In addition, metallic alloying can also be a promising means. In the latter, the dielectric function of the particles no longer obeys the Drude

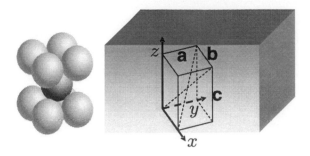

Figure 3.4. Schematic of the colloidal crystals immersed in a graded host medium. $a = b$ is assumed in the numerical calculations, which forms square dipolar lattice in xy-plane. After Ref. [50].

form, and the theoretical calculations are formidable because it involves a Green function formalism for band structure, and a linear response theory (i.e., Kubo formula) for transport properties [48]. In this section, the objective of considering tetragonal lattices is to achieve anisotropy by changing the lattice parameters, in order to produce a larger optical nonlinearity than in an isotropic system such as a cubic structure.

In summary, we theoretically exploit a class of nonlinear optical materials based on colloidal crystals of graded metallodielectric nanoparticles. Such materials can have both an enhancement and a red shift of optical nonlinearity, due to the gradation inside the metallic core as well as the lattice effects arising from the periodic structure.

2.   With a Graded-Index Host

We shall theoretically investigate a colloidal crystal immersed in a graded-index host and demonstrate a giant enhanced optical nonlinearity band, which is controllable by the gradient and by the easily-tunable colloid structure as well. We basically used the quasistatic point-dipole approximation, which suffices in terms of characterizing both the gradient effects and the lattice effects, otherwise the solution is formidable, either from a Green's function formalism or from first-principles [49].

The theoretical calculations are deployed on a model tetragonal lattice with uniaxial anisotropy [see Fig. 3.4] [40]. Without loss of generality, we will only discuss the bct, bcc and the fcc lattices respectively. The bcc has the lowest packing density while the fcc has the highest one among the three cases [40]. Extensions to other colloidal structures such as the simple tetragonal lattice are straightforward and similar results are expected. Taking advantages of the interlayer interaction tensor $\mathbf{T}$ (i.e., T_{ij} denotes the interaction strength between two in-plane dipole arrays) given by the Lekner summation method, [51, 52, 53] we solved the self-consistent equations

$$\mathbf{E}_i = \frac{1}{a^3} \sum_{j=1}^{N} \mathbf{T}_{ij} \cdot (\alpha_j \mathbf{E}_j) + \mathbf{E}_i^{(0)}, \qquad (3.25)$$

where a is the lattice constant as shown in Fig. 3.4, α_j is the layer-dependent linear bare polarizability, here i, j label the crystal layer and N denotes the total layer number. $\mathbf{E}_i^{(0)}$ in the self-consistent equation is not simply the externally applied electric field $\mathbf{E}_0$ due to the presence of gradient. However, it is the field inside the graded host medium, which is thus determined by virtue of the continuity of the normal component of the electric displacement $\mathbf{D}$ in the longitudinal case, i.e., $\mathbf{E}_0$ parallel to the uniaxial axis. It is the z-axis as shown in Fig. 3.4 in our case. Nevertheless in the transverse case ($\mathbf{E}_0$ perpendicular to the uniaxial axis), we exactly used the applied field $\mathbf{E}_0$ because the boundary condition now becomes the continuity of the tangential component of electric field. We compared the effective linear and nonlinear optical responses of colloidal crystals with the different lattice structures (i.e., bct, bcc and fcc), made of metallic nanoparticles of linear dielectric constant ε_1 and third-order nonlinear susceptibility χ_1, suspended in a host fluid of ε_m [see Fig. 3.4]. Both the longitudinal (L) and transverse (T) results will be discussed. The self-consistent equations over $i = 1, 2 \cdots, N$ are then combined together to take into account the lattice effect and is being able to be transformed into a matrix form as

$$\mathbf{E} = \mathbf{TAE} + \mathbf{E}^{(0)}. \tag{3.26}$$

More precisely, in the longitudinal and the transverse cases, $\mathbf{E} = \{E_i^{(L,T)}\}$ is simply N-dimensional vector and $\mathbf{A}$ is $N \times N$ diagonal matrix of the polarizability, which relates the induced dipole moment of the particle in the layer ℓ_i and the local field $\mathbf{E}_i$, and indeed consists of the linear and nonlinear contributions. That is

$$\mathbf{p}_i = \alpha_i \mathbf{E}_i + \beta_i |\mathbf{E}_i|^2 \mathbf{E}_i / 3, \tag{3.27}$$

where $\alpha_i = \varepsilon_m r^3 (\varepsilon_1 - \varepsilon_m)/(\varepsilon_1 + 2\varepsilon_m)$. Here r is the radii of the metallic nanoparticles. In view of weak nonlinearity in the colloidal particles with the nonlinear relationship $\mathbf{D} = \varepsilon_1 \mathbf{E} + \chi_1 |\mathbf{E}|^2 \mathbf{E}$, by using the perturbation expansion method, [54] we obtain

$$\beta_i = \left(\frac{3\varepsilon_m}{\varepsilon_1 + 2\varepsilon_m} \right)^2 \left| \frac{3\varepsilon_m}{\varepsilon_1 + 2\varepsilon_m} \right|^2 r^3 \chi_1. \tag{3.28}$$

It is noteworthy that the linear local field $\mathbf{E}_i$ around the particles in the layer ℓ_i are actually obtained by assuming no intrinsic nonlinear response, i.e., we set $\chi_1 = 0$ for solving the self-consistent equations, which is appropriate provided that the nonlinear responses are much less than the linear ones. Next we use the resultant linear local fields $\mathbf{E}_i$ to extract the enhancement factor of the effective nonlinear susceptibility [54, 55]

$$\gamma = \frac{\bar{\chi}_1}{\chi_1} = \rho \frac{\langle |E_i|^2 E_i^2 \beta_i \rangle}{3 E_0^4 \chi_1}, \tag{3.29}$$

where $\rho = \pi[(q^3+2)/q]^{3/2}/24$ is the total volume fraction of the metallic colloidal nanoparticles. $1-q$ quantifies the degree of anisotropy of the periodic lattice [40], which also determines the interlayer interaction $\mathbf{T}$, thus results in structure-controllable optical properties. Note that the averaging $\langle ... \rangle$ in Eq. (3.29) is taken over the layers $\ell_i (i = 1, \cdots, N)$ instead of over the nanoparticles spatial volume, because in this approximation the local fields inside each of the particles are homogeneous. We also assume no nonlinear response in the host, which is in fact relatively neglectable comparing to that in the metal. Additionally, a gradient of the dielectric constant of the host fluid is introduced along the uniaxial direction

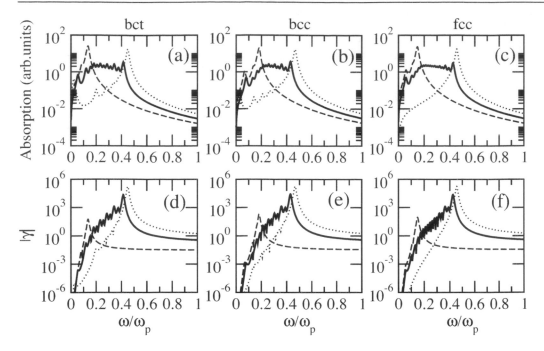

Figure 3.5. The effective linear absorption and third-order nonlinearity enhancement factor of the periodic colloid nanoparticles (diameter $a = 1$) immersed in a graded host fluid whose dielectric constant varies as a function of the layer index i as $\varepsilon_2(i) = 1 + 1.25i/N$, and in homogeneous host fluids with dielectric constant $\varepsilon_2 = 1.5^2$ (dashed) and $\varepsilon_2 = 1.0$ (dotted), respectively. The dielectric function of the metallic colloidal nanoparticles are simply denoted by the Drude form $\varepsilon_1 = 1 - \omega_p^2/(\omega^2 + i\omega\Gamma)$. Parameters: $\Gamma = 0.02\omega_p$, $N = 25$, $r = 1$, and $E_0 = 1$. After Ref. [50].

of the colloidal crystallines, i.e, $\varepsilon_m = \varepsilon_m(z_i)$ in our case. In this regard, we treat the host as a continuously-layered film, thus one explicitly has $\mathbf{E}_0 = \{E_i^{(0)} = E_0/\varepsilon_m(z_i)\}$ in the longitudinal case. The formation of the gradient in the host might be achieved by dispersing different polymers in it, by selectively filling with microfluidic materials, [56] or induced by the presence of a temperature gradient, etc. One can also simply coat the nanoparticles with different coverage shells. But it still remains a challenge because the novel properties from our prediction require a reasonably large gradient in the dielectric constant of the host.

Figure 3.5 shows in logarithmic scale the longitudinal optical absorption and the modulus of the nonlinearity enhancement factor γ defined in Eq. (3.29), as functions of the reduced frequency [see the figure caption for more details]. We specifically compared the results of bct ($q = 0.87358$), bcc ($q = 1.0$) and fcc ($q = 1.25992$) as shown in the three columns. The presence of the inhomogeneity in the host fluid obviously leads to a broadened and giant enhanced resonant band in the low-frequency region. This is interesting for potential telecommunication applications. The results of the same colloid suspended in homogeneous host medium with $\varepsilon_m = 1$ (dotted-lines) and $\varepsilon_m = 2.25$ (dashed-lines) are also presented, in an attempt to demonstrate that the broadened resonant band in some sense stems from the hybridization of the non-graded structures. From the absorption spectrum

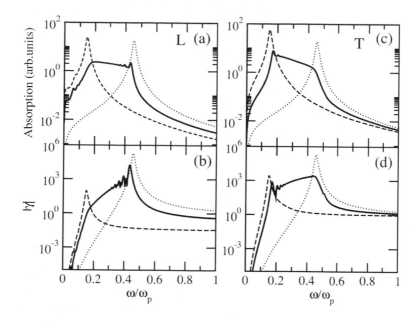

Figure 3.6. The same as Fig. 3.5, but with totally $N = 50$ layers in the fcc lattice. (a) and (b) the longitudinal case; (c) and (d) the transverse case. After Ref. [50].

and the enhancement in the third-order nonlinear susceptibility, we would expect an at tractive figure of merit. [12, 13, 14, 15, 16] That is, the materials effectively exhibit large nonlinearity and relatively small absorption. This is certainly superior to pure metal be- cause it generally has large nonlinearity and unwanted absorption concomitantly. In fact, the optical absorption arises from the surface plasmon resonance, which is obtained from the imaginary part of the effective dielectric constant that is extracted from the generalized Clausius-Mossotti formula. [40, 49] Note that the plasmon resonant peaks in the cases of homogeneous host fluid are redshifted with respect to the corresponding ones [not shown] predicted form the Maxwell-Garnett theory. We actually set a gap (e.g., the coated layer thickness in experiments) of $2r/5$ between the nearest lattice particles in order to avoid the severe complications arising from the multiple image interactions. It is well known that the multipolar interactions play crucial roles when two particles approach. [49] The introduc- tion of the gap indeed makes the nanoparticles size, and thus the dipole factor relatively small and somewhat unfavorably suppresses the effect arising from variation of the lattice structure, as seen in Fig. 3.5.

Furthermore, due to the fact that we treated the continuous variations of dielectric func- tion in the host as layered ones in obtaining α_i and β_i, and the fact that the dipoles are actually distributed in discrete lattice nodes, a series of sharp peaks are also observable in Fig. 3.5. The peaks are merged in the broadened band and they become more notable for a increased gradient in the host dielectric constant [not shown], whereas they tend to dis- appear as the crystal layer N is increased. This is also understandable in the generalized Bergman-Milton spectral representation in graded composites [57]. In detail, the merging is explicitly shown in Fig. 3.6, where we increased the layer to $N = 50$. The fcc is taken as an example and we present both the longitudinal and the transverse results. The peaks in Fig. 3.5(c) and Fig. 3.5(f) are distinctly smeared out in Fig. 3.6(a) and Fig. 3.6(b) respectively. The transverse results [see Fig. 3.6(c) and (d)] in the presence of the gradient is slightly different to that in the longitudinal case, but still retain the broadened bands. We ascribe this to the fact that the layer-to-layer interactions fall off exponentially due to the screening effect from the lattice, [51, 52, 53] therefore give no much layer-structure-dependent interactions for both the two cases. Note that the longitudinal and transverse results of crystals in homogeneous host [dotted and dashed lines in Fig. 3.6] do not differ much as well.

In conclusion, we theoretically exploit the optical resonant enhancement due to lattice effect and gradient effect in colloidal crystals, which are made out of suspended metallic nanoparticles in a graded-index host. The gradient in the fluid and the colloid structure are easily subjected to tunability, for example, the structure transformation might be induced by electrorheological effects or by self-assembly of two kinds of particles with biochemically different surface properties, etc. In addition, one can also use metal-covered mangetic nanoparticles and control the suspension structure by external magnetic field, consequently realizing magneto-controlled optical properties [7]. In this case, the electro-magnetorheological effects is a good candidate as well. Devices that could benefit from these materials include optical switches, optical limiters, as well as biosensors, etc.

Chapter 4

Inhomogeneous Metallic Films

Thin films can possess different optical properties (see, e.g., Ref. [58]) when comparing with their bulk counterparts. Recently, some authors found experimentally that the graded thin films may have better dielectric properties than a single-layer film [59]. Owing to the additional control of gradation, graded thin films play a key role in many applications and hence are essential for material fabrication. Graded materials [43] are the materials whose material properties can vary continuously in space. These materials have attracted much interest in various engineering applications [60]. When compared to bulk materials, the corresponding thin films often possess different optical properties [58, 61]. However, the traditional theories [62] cannot deal with the composites of graded inclusions directly. Recently, for treating these composites, we presented a first-principles approach [12, 63] and a differential effective dipole approximation [64, 65].

The problem becomes more complicated by the presence of nonlinearity in realistic composites. Besides gradation (inhomogeneity), nonlinearity plays also an important role in the effective material properties of composite media [10, 55, 66, 67, 68, 69, 70, 71, 72, 73, 74]. A large nonlinearity enhancement was found indeed when the authors studied a sub-wavelength multilayer of titanium dioxide and conjugated polymer [4]. For nonlinear effects other than the Kerr effect, Hui *et al.* [75] derived general expressions for the effective susceptibility for the second-harmonic generation (SHG) in a binary composite of random dielectrics. They have also studied the thickness dependence of effective SHG susceptibility in films of random dielectrics and in composites with coated small particles [76, 77]. Until now, achieving enhanced SHG has been a challenge [78, 79]. For instance, recently one obtained enhanced SHG by means of resonant waveguide gratings incorporating ionic self-assembled monolayer polymer films [79]. In addition, nondegenerate four-wave mixing [80, 81, 82, 83] has also received much attention due to the abundant application in phase conjugate mirrors, wave restoration, real time holography, etc.

As a matter of fact, in practice it is more convenient to fabricate multilayer metallic films than graded films as multilayer metallic films can be readily prepared in a filtered arc deposition system. Therefore, it is necessary to discuss the multilayer effect as the number of layers inside the films increases. In this regard, this should be expected to have practical relevance. As the number of layers N increases, we shall show a gradual transition from sharp peaks to an emerging broad continuous band and the graded film results recover in the limit $N \to \infty$.

1. Graded Metallic Films

1.1. Third-Order Nonlinearity

Let us consider a graded metallic film with width L, and the gradation under consideration is in the direction perpendicular to the film. The local constitutive relation between the displacement ($\mathbf{D}$) and the electric field ($\mathbf{E}$) inside the graded layered geometry is given by

$$\mathbf{D}(z,\omega) = \varepsilon(z,\omega)\mathbf{E}(z,\omega) + \chi(z,\omega)|\mathbf{E}(z,\omega)|^2\mathbf{E}(z,\omega), \tag{4.1}$$

where $\varepsilon(z,\omega)$ and $\chi(z,\omega)$ are respectively the linear dielectric constant and third-order nonlinear susceptibility. Note that both $\varepsilon(z,\omega)$ and $\chi(z,\omega)$ are gradation profiles as a function of position r. Here we assume that the weak nonlinearity condition is satisfied, that is, the contribution of the second term (nonlinear part $\chi(z,\omega)|\mathbf{E}(z,\omega)|^2$) in the right-hand side of Eq. (4.1) is much less than that of the first term (linear part $\varepsilon(z,\omega)$) [66]. We further restrict our discussion to the quasi-static approximation, under which the whole layered geometry can be regarded as an effective homogeneous one with effective (overall) linear dielectric constant $\bar{\varepsilon}(\omega)$ and effective (overall) third-order nonlinear susceptibility $\bar{\chi}(\omega)$. To show the definitions of $\bar{\varepsilon}(\omega)$ and $\bar{\chi}(\omega)$, we have [66]

$$\langle \mathbf{D} \rangle = \bar{\varepsilon}(\omega)\mathbf{E}_0 + \bar{\chi}(\omega)|\mathbf{E}_0|^2\mathbf{E}_0, \tag{4.2}$$

where $\langle \cdots \rangle$ denotes the spatial average, and $\mathbf{E}_0 = E_0\hat{e}_z$ is the applied field along the $z-$axis.
   We adopt the graded Drude dielectric profile

$$\varepsilon(z,\omega) = 1 - \frac{\omega_p^2(z)}{\omega(\omega + i\gamma(z))}, \quad 0 \leq z \leq L. \tag{4.3}$$

In Eq.(5.23), we adopted various plasma-frequency gradation profile

$$\omega_p(z) = \omega_p(0)(1 - C_\omega \cdot z/L), \tag{4.4}$$

and relaxation-rate gradation profile [84]

$$\gamma(z) = \gamma(\infty) + \frac{C_\gamma}{z/L}, \tag{4.5}$$

where C_ω is a dimensionless constant (gradient). Here $\gamma(\infty)$ denotes the damping coefficient in the corresponding bulk material. C_γ is a constant (gradient) which is related to the Fermi velocity. A z-dependent profile for the plasma frequency and the relaxation rate can be achieved experimentally. One possible way may be to impose a temperature profile, as it has been observed that surface enhanced Raman scattering is sensitive to temperature [85]. Thus, one may tune the surface plasmon frequency by imposing an appropriate temperature gradient [86]. A temperature gradient may also be used in materials with a small band gap or with a profile on dopant concentrations. In this case, one may impose a charge carrier concentration profile to a certain extent. This effect, when coupled with materials with a significant intrinsic nonlinear susceptibility, will give us with a way to control the effective nonlinear response. For less conducting materials, one may replace the Drude form of dielectric constants by a Lorentz oscillator form. It may also be possible to fabricate dirty metal films in which the degree of disorder varies in the z-direction and hence leads to a relaxation-rate gradation profile. Nevertheless, the present results do not depend crucially on the particular form of the dielectric function. The only requirement is that we must have a sufficiently large gradient, either in $\omega_p(z)$ or in $\gamma(z)$ to yield a broad plasmon band.

Due to the simple layered geometry, we can use the equivalent capacitance of series combination to calculate the linear response, i.e., the optical absorption for the metallic film:

$$\frac{1}{\bar{\varepsilon}(\omega)} = \frac{1}{L} \int_0^L \frac{dz}{\varepsilon(z,\omega)}. \tag{4.6}$$

The calculation of nonlinear optical response can proceed as follows. We first calculate local electric field $E(z,\omega)$ by the identity

$$\varepsilon(z,\omega)E(z,\omega) = \bar{\varepsilon}(\omega)E_0$$

by virtue of the continuity of electric displacement, where E_0 is the applied field.

In view of the existence of nonlinearity inside the graded film, the effective nonlinear response $\bar{\chi}(\omega)$ can be written as [66]

$$\bar{\chi}(\omega)\mathbf{E}_0^4 = \langle \chi(z,\omega)|\mathbf{E}_{\mathrm{lin}}(z)|^2 \mathbf{E}_{\mathrm{lin}}(z)^2 \rangle, \tag{4.7}$$

where E_{lin} is the linear local electric field. Next, the effective nonlinear response can be written as an integral over the layer such as

$$\bar{\chi}(\omega) = \frac{1}{L} \int_0^L dz \chi(z,\omega) \left| \frac{\bar{\varepsilon}(\omega)}{\varepsilon(z,\omega)} \right|^2 \left(\frac{\bar{\varepsilon}(\omega)}{\varepsilon(z,\omega)} \right)^2. \tag{4.8}$$

For numerical calculations, we set $\chi(z,\omega)$ to be constant (χ_1), in an attempt to emphasize the enhancement of the optical nonlinearity. Without loss of generality, the layer width L is taken to be 1.

Figure 4.1 displays the optical absorption $\sim \mathrm{Im}[\bar{\varepsilon}(\omega)]$, the modulus of the effective third-order optical nonlinearity enhancement $|\bar{\chi}(\omega)|/\chi_1$, as well as the FOM $|\bar{\chi}(\omega)|/\{\chi_1 \mathrm{Im}[\bar{\varepsilon}(\omega)]\}$ as a function of the incident angular frequency ω. Here $\mathrm{Im}[\cdots]$ means the imaginary part of $\cdots$. To one's interest, when the positional dependence of $\omega_p(z)$ is taken into account (namely, $C_\omega \neq 0$), a broad resonant plasmon band is observed. As expected, the broad band is caused to appear by the effect of the positional dependence of the plasma frequency of the graded metallic film. In particular, this band can be observed within almost the whole range of frequency, as the gradient C_ω is large enough. In other words, as long as the film under consideration is strongly inhomogeneous, a resonant plasmon band is expected to appear over the whole range of frequency. In addition, it is also shown that increasing C_ω causes the resonant bands to be red-shifted (namely, located at a lower frequency region). In a word, although the enhancement of the effective third-order optical nonlinearity is often accompanied with the appearance of the optical absorption, the FOM is still possible to be quite attractive due to the presence of the gradation of the metallic film.

Similarly, in Figure 4.2, we investigate the effect of the inhomogeneity of the relaxation rates [$\gamma(z)$], which comes from the graded metallic film. It is evident to show that, in the

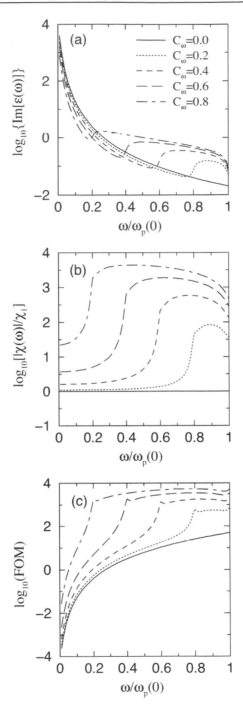

Figure 4.1.   (a) The linear optical absorption $\mathrm{Im}[\bar{\varepsilon}(\omega)]$, (b) the enhancement of the third-order optical nonlinearity $|\bar{\chi}(\omega)|/\chi_1$, and (c) the FOM (figure of merit) $\equiv |\bar{\chi}(\omega)|/\{\chi_1\mathrm{Im}[\bar{\varepsilon}(\omega)]\}$ versus the normalized incident angular frequency $\omega/\omega_p(0)$ for dielectric function gradation profile $\varepsilon(z,\omega) = 1 - \omega_p^2(z)/[\omega(\omega + i\gamma(z))]$ with various plasma-frequency gradation profile $\omega_p(z) = \omega_p(0)(1 - C_\omega \cdot z/L)$ and relaxation-rate gradation profile $\gamma(z) = \gamma(\infty) + \frac{C_\gamma}{z/L}$. Parameters: $\gamma(\infty) = 0.02\omega_p(0)$ and $C_\gamma = 0.0$. After Ref. [14].

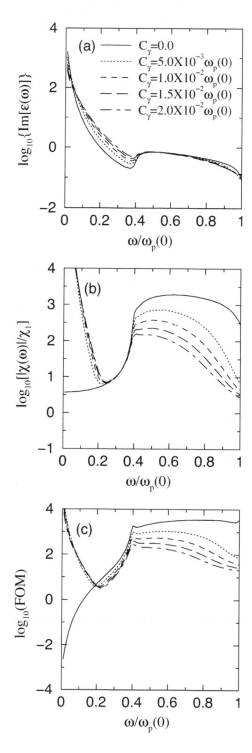

Figure 4.2. Same as Fig. 4.1. Parameters: $\gamma(\infty) = 0.02\omega_p(0)$ and $C_\omega = 0.6$. After Ref. [14].

low-frequency region, the positional dependence of relaxation rate $\gamma(z)$ enhances not only the third-order optical nonlinearity but also the FOM of such kind of graded metallic films.

Consequently, graded metallic films can be a suitable candidate material for obtaining the *optimal* FOM. Thus, corresponding experiments are expected to be done to check our theoretical predictions since graded films can be fabricated easily.

We have discussed a graded metallic film (layered geometry), in an attempt to investigate the effect of gradation on the nonlinear enhancement and FOM of such materials. It should be remarked that the optical response of the layered geometry depends on polarization of the incident light, because the incident optical field can always be resolved into two polarizations. However, a large nonlinearity enhancement occurs only when the electric field is parallel to the direction of the gradient [4], and the other polarization does not give nonlinearity enhancement at all [4].

In the conventional theory of surface plasmon resonant nonlinearity enhancement, there is often a dielectric component in the system of interest. In this regard, it turns out that it is not difficult to add a homogeneous dielectric layer on the metallic film. The same theory still works but a prominent surface plasmon resonant peak appears at somewhat lower frequencies in addition to the surface plasmon band. Due to the concomitantly strong absorption, the figure of merit of the resonant enhancement peak is too small to be useful. In the limit of vanishing volume fraction of the dielectric component, however, the present results recover.

To sum up, we have investigated the effective linear and third-order nonlinear susceptibility of graded metallic films with weak nonlinearity. We calculated the effective linear dielectric constant and third-order nonlinear susceptibility. It has been found that the presence of gradation in metallic films yields a broad resonant plasmon band in the optical region, resulting in a large nonlinearity enhancement and hence an optimal FOM.

1.2. Second-Harmonic Generation

Let us consider a graded metallic film with thickness L, with the gradation profile in the direction perpendicular to the film. If we only include quadratic nonlinearities, the local constitutive relation between the displacement field $\mathbf{D}(z)$ and the electric field $\mathbf{E}(z)$ in the static case is [76, 77] $D_i(z) = \sum_j \varepsilon_{ij}(z)E_j(z) + \sum_{jk}\chi_{ijk}(z)E_j(z)E_k(z)$ with $i = x, y, z$, where $D_i(z)$ and $E_i(z)$ are the ith component of $\mathbf{D}(z)$ and $\mathbf{E}(z)$, respectively, and χ_{ijk} is the SHG susceptibility. Here $\varepsilon_{ij}(z)$ denotes the linear dielectric function, which we assume for simplicity to be isotropic $\varepsilon_{ij}(z) = \varepsilon(z)\delta_{ij}$. Both $\varepsilon(z)$ and $\chi_{ijk}(z)$ are functions of z and describe the gradation profiles.

In general, when a monochromatic external field is applied, the nonlinearity will generate local potentials and fields at all harmonic frequency. For a finite frequency external electric field of the form $E_0 = E_0(\omega)e^{-i\omega t} + \text{c.c.}$, the effective SHG susceptibility $\bar{\chi}_{2\omega}$ can be extracted by considering the volume average of the displacement field at the frequency 2ω in the inhomogeneous medium [75, 76, 77].

Here, we adopt a graded dielectric profile that follows the Drude form

$$\varepsilon(z, \omega) = 1 - \frac{\omega_p^2(z)}{\omega(\omega + i\gamma(z))}, \tag{4.9}$$

where $0 \leq z \leq L$. The general form in Eq. (4.9) allows for the possibility of a gradation profile in the plasma frequency [e.g., Eq. (4.29)] and the relaxation rate [e.g., Eq. (4.5)]. For a z-dependent profile, we can make use of the equivalent capacitance for capacitors in series to evaluate the effective perpendicular linear response for a given frequency, i.e., $1/\bar{\varepsilon}(\omega) = (1/L) \int_0^L dz[1/\varepsilon(z,\omega)]$. Using the continuity of the electrical displacement field, the local electric fields $E(z,\omega)$ satisfies

$$\varepsilon(z,\omega)E(z,\omega) = \bar{\varepsilon}(\omega)E_0(\omega), \qquad (4.10)$$

where $E_0(\omega)$ is the applied field along z axis.

The calculation of the effective nonlinear optical response then proceed by applying the expressions derived in Refs. [76, 77]. Next, the effective SHG susceptibility $\bar{\chi}_{2\omega}$ is given by $\bar{\chi}_{2\omega} = \langle \chi_{2\omega}(z)E_{\text{lin}}(z,2\omega)E_{\text{lin}}(z,\omega)^2 \rangle / [E_0(2\omega)E_0(\omega)^2]$, where E_{lin} is the linear local electric field in the graded film with the same gradation profile but with a vanishing nonlinear response at the frequency concerned. Using Eq. (4.10) for the linear local fields, the effective SHG susceptibility can be rewritten as an integral over the film as

$$\bar{\chi}_{2\omega} = \frac{1}{L} \int_0^L dz \chi_{2\omega}(z) \left(\frac{\bar{\varepsilon}(2\omega)}{\varepsilon(z,2\omega)} \right) \left(\frac{\bar{\varepsilon}(\omega)}{\varepsilon(z,\omega)} \right)^2. \qquad (4.11)$$

To illustrate the SHG in graded films, we consider as a model system that the intrinsic SHG susceptibility $\chi_{2\omega}(z) = \chi_1$ to be a real and positive frequency-independent constant and does not have a gradation profile. In doing so, we are allowed to focus on the enhancement of the SHG response when compared with χ_1. It should be noted that the present results will not depend crucially on the particular form of the gradation profile. The point is that one needs a sufficiently large gradient. To show the effects of gradation, here we take as a model plasma-frequency gradation profile

$$\omega_p(z) = \omega_p(0)(1 - C_\omega \cdot z), \qquad (4.12)$$

and a model relaxation-rate gradation profile [84]

$$\gamma(z) = \gamma(\infty) + C_\gamma/z, \qquad (4.13)$$

where C_ω and C_γ are constant parameters tuning the profile which is assumed to be linear in z. Here $\gamma(\infty)$ denotes the bulk damping coefficient, i.e., for $z \to \infty$. Set thickness $L = 1$ so that we could focus on the film with a fixed thickness. Regarding the thickness dependence, we refer the reader to the work by Hui et al. [76]

Figure 4.3 shows the real and imaginary parts of the effective linear dielectric constant [Fig. 4.3(a) and Fig. 4.3(b)], and the real and imaginary parts of the effective SHG susceptibility [Fig. 4.3(c) and Fig. 4.3(d)] as a function of frequency $\omega/\omega_p(0)$. Also shown is the modulus of $\bar{\chi}_{2\omega}/\chi_1$, see Fig. 4.3(e). The dielectric function gradation profile is given in Eqs. (4.9), (4.12) and (4.13) with $C_\gamma = 0$, i.e., only a graded plasmon frequency is included. Throughout the calculations, the real part of the (linear) dielectric constant is negative naturally. In this case, a broad resonant plasmon band is observed. Note that for $C_\omega \to 0$, $\omega_p(z)/\omega_p(0) \to 1$. As C_ω increases, $\omega_p(z)$ takes on values within a broader range across the film, and leads to a broad plasmon band. Increasing C_ω also leads to a shift to the plasmon peak to lower frequencies. The reason is that, in analogous to capacitors in series, the

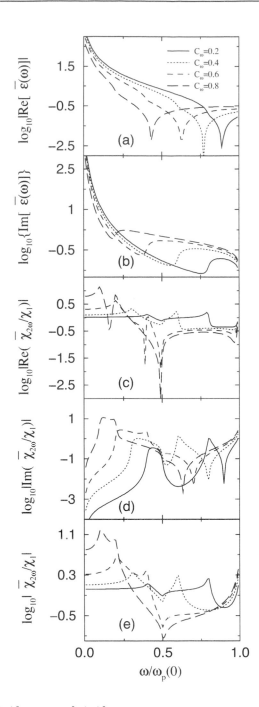

Figure 4.3. (a) $\mathrm{Re}[\bar{\varepsilon}(\omega)]$, (b) $\mathrm{Im}[\bar{\varepsilon}(\omega)]$ (linear optical absorption), (c) $\mathrm{Re}[\bar{\chi}_{2\omega}/\chi_1]$, (d) $\mathrm{Im}[\bar{\chi}_{2\omega}/\chi_1]$, and (e) Modulus of $\bar{\chi}_{2\omega}/\chi_1$ versus the normalized incident angular frequency $\omega/\omega_p(0)$ for the dielectric function gradation profile [Eq. (4.9)] with various plasma-frequency gradation profile [Eq. (4.12)] and relaxation-rate gradation profile [Eq. (4.13)]. Here $|\cdots|$ denotes the absolute value or modulus of $\cdots$. Parameters: $\gamma(\infty) = 0.02\omega_p(0)$ and $C_\gamma = 0.0$. After Ref. [16].

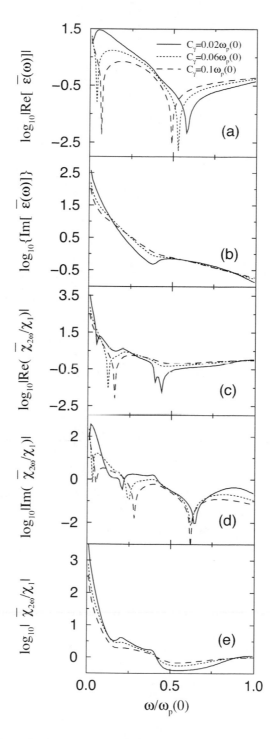

Figure 4.4. Same as Fig. 4.3. Parameters: $\gamma(\infty) = 0.02\omega_p(0)$ and $C_\omega = 0.6$. After Ref. [16].

effective dielectric constant of the film is dominated by the layer with the smallest dielectric constant. For the SHG response, the frequency dependence is quite complicated. As C_ω increases, it is noted that structures in the SHG response also shifts to the lower frequencies and the range of values of the SHG susceptibility increases as well.

Figure 4.4 displays the results of model calculations in which a gradation profile of the relaxation rate of the form in Eq. (4.13) is also included. The effects are similar to those in Fig. 4.3. The SHG response is found to be enhanced more strongly in the presence of both a relaxation-rate gradation and a plasma-frequency gradation (see Fig. 4.3) than for plasmon-frequency gradation alone, especially at low frequencies. As C_γ increases, the structures in the linear and SHG response both show a shift to lower frequencies.

In Figs. 4.3-4.4, the quantities which can be both positive and negative are plotted in logarithm of modulus. When the quatitites pass through zero, the logarithm will be very large, thus yielding spikes. In addition, we have used the normalized numbers in order to describe a general origin of the SHG in metal films rather than a specified metal film.

Again, the point for achieving the present results is that one needs a sufficiently large gradient rather than a crucially particular form of the dielectric function or gradation profiles.

In summary, we have carried out model numerical calculations for the effective SHG susceptibility in a graded metallic films, with linear gradation profile possibly imposed on the plasma frequency and the relaxation time. It is found that the presence of a gradation profile leads to a broader structure in both the linear and SHG response and an enhancement in the SHG signal. Therefore, the graded metallic films can serve as a novel optical material.

1.3. Nondegenerate Four-Wave Mixing

In the particular NDFM (Nondegenerate four-wave mixing) of our interest, there are four waves: Forward pump wave E_f and backward pump wave E_b at angular frequency ω_1 (propagating in opposite directions); probe wave E_p and signal wave E_s at ω_2. We shall impose the two pump fields (incident field) at ω_1 of high intensity to generate the desired nonlinearity, while the probe field at ω_2 of lower intensity is measured. This is experimentally observable. When $\omega_2 = \omega_1$, one has degenerate four-wave mixing susceptibility, which is exactly the same as the usual effective third-order nonlinearity susceptibility $\bar{\chi}^{(3)}(\omega_2)$, namely,

$$\bar{\chi}^{(3)}(\omega_2) = \langle \chi(\omega_2) F(\omega_2)^2 |F(\omega_2)|^2 \rangle, \tag{4.14}$$

where F denotes the local-field enhancement factor, $\chi(\omega_2)$ the (intrinsic) third-order nonlinear susceptibility, and $\langle \cdots \rangle$ the volume average of $\cdots$. For the present effective NDFM susceptibility $\bar{\chi}(\omega_2)$, we obtain the general form such that [2, 73, 87]

$$\bar{\chi}(\omega_2) = \langle \chi(\omega_2) F(\omega_2)^2 |F(\omega_1)|^2 \rangle. \tag{4.15}$$

It is worth remarking that all the above-mentioned E-fields can make an angle to the gradient which is directed along z axis. However, only the components that are parallel to z axis give rise to a significant enhancement while the other components that are perpendicular to z axis lead to no enhancement [4]. Thus, the third-order nonlinear susceptibility plays a crucial role in the effective NDFM susceptibility, as given by Eq. (6.1).

Let us consider a graded metallic film of width L. Its gradation is directed along z axis, i.e., perpendicular to the film. For each layer at position z inside the graded film, the local constitutive relation between the displacement $\mathbf{D}(z,\omega)$ and the electric field $\mathbf{E}(z,\omega)$ is given by

$$\mathbf{D}(z,\omega) = \varepsilon(z,\omega)\mathbf{E}(z,\omega) + \chi(z,\omega)|\mathbf{E}(z,\omega)|^2\mathbf{E}(z,\omega), \qquad (4.16)$$

where $\varepsilon(z,\omega)$ and $\chi(z,\omega)$ stand for the linear dielectric constant and third-order nonlinear susceptibility, respectively. Here both $\varepsilon(z,\omega)$ and $\chi(z,\omega)$ are gradation profiles as a function of position z, and ω denotes the angular frequency of the fields, which will be distinguished in the following, see Eq. (4.21). Next, we again assume that the weak nonlinearity condition is satisfied. We further focus on the quasi-static approximation, under which the whole graded film can be regarded as an effective homogeneous one with effective (overall) linear dielectric constant $\bar{\varepsilon}(\omega)$ and effective (overall) third-order nonlinear susceptibility $\bar{\chi}(\omega)$. They both satisfy the following definition [66]

$$\langle\mathbf{D}\rangle = \bar{\varepsilon}(\omega)\mathbf{E}_0 + \bar{\chi}(\omega)|\mathbf{E}_0|^2\mathbf{E}_0, \qquad (4.17)$$

where $\mathbf{E}_0 = E_0\hat{e}_z$ is the applied field along $z-$axis.

Then, we adopt the graded Drude dielectric profile

$$\varepsilon(z,\omega) = 1 - \frac{\omega_p^2(z)}{\omega[\omega + i\gamma(z)]}, \quad z \leq L, \qquad (4.18)$$

where $\omega_p(z)$ and $\gamma(z)$ represent the plasma-frequency gradation profile and relaxation-rate gradation profile, respectively, and are both as a function of z.

In view of the present graded film (layered geometry) of interest, we are allowed to use the equivalent capacitance of series combination to calculate the effective linear dielectric constant $\bar{\varepsilon}(\omega)$ for the metallic film,

$$\frac{1}{\bar{\varepsilon}(\omega)} = \frac{1}{L}\int_0^L \frac{dz}{\varepsilon(z,\omega)}. \qquad (4.19)$$

Next, we are in a position to derive the nonlinear optical response. First, we calculate local electric field $E(z,\omega)$ by using the identity

$$\varepsilon(z,\omega)E(z,\omega) = \bar{\varepsilon}(\omega)E_0, \qquad (4.20)$$

which arises from the continuity of the electric displacements. In view of the existence of nonlinearity inside the graded film, based on Eq. (6.1), the effective NDFM susceptibility $\bar{\chi}(\omega_2)$ can be rewritten as an integral over the whole graded film such that

$$\bar{\chi}(\omega_2) = \frac{1}{L}\int_0^L dz\,\chi(z,\omega_2)\left(\frac{\bar{\varepsilon}(\omega_2)}{\varepsilon(z,\omega_2)}\right)^2\left|\frac{\bar{\varepsilon}(\omega_1)}{\varepsilon(z,\omega_1)}\right|^2. \qquad (4.21)$$

To illustrate the NDFM response in graded films, we consider as a model system in which the intrinsic NDFM susceptibility $\chi(z,\omega_2)$ is assumed to be a real and positive frequency-independent constant χ_1 and does not have a gradation profile either. By doing so, we could focus on the enhancement of the NDFM response. Without loss of generality, the film thickness L is set to be unity. For numerical calculations, we adopt the plasma-frequency gradation profile

$$\omega_p(z) = \omega_p(0)(1 - C_\omega \cdot z), \tag{4.22}$$

and relaxation-rate gradation profile [84]

$$\gamma(z) = \gamma(\infty) + C_\gamma/z, \tag{4.23}$$

where C_ω is a constant tuning the profile. Here $\gamma(\infty)$ denotes the damping coefficient in the corresponding bulk material, and C_γ is a constant which is related to the Fermi velocity.

Figure 4.5 displays (a) the linear optical absorption $\text{Im}[\bar{\varepsilon}(\omega_2)]$, (b) the enhancement of the NDFM susceptibility $|\bar{\chi}(\omega_2)|/\chi_1$, and (c) the FOM $\equiv |\bar{\chi}(\omega_2)|/\{\chi_1 \text{Im}[\bar{\varepsilon}(\omega_2)]\}$ versus the normalized incident angular frequency $\omega_2/\omega_p(0)$, for various C_ω. The dielectric function gradation profile is given in Eqs. (4.18), (4.22), and (4.23) with $C_\gamma = 0$, i.e., only a graded plasmon frequency is included. To one's interest, when the plasmon-frequency gradation profile $\omega_p(z)$ is taken into account, a broad resonant plasmon band is observed always. In other words, the broad band is caused to appear by the effect of the gradation. In detail, as $C_\omega \to 0$, $\omega_p(z)/\omega_p(0) \to 1$. As C_ω increases, $\omega_p(z)$ has values within a broad range, thus yielding a broad plasmon band. Moreover, we find that increasing C_ω causes the resonant band not only to be enhanced, but also red-shifted (namely, located at a lower frequency region). The reason is that, in analogous to capacitors in series, the effective linear dielectric constant of the graded film (i.e., multilayer structure) is dominated by the layer with the smallest dielectric constant. From Fig. 4.5(b), we find that increasing C_ω causes the peak of the NDFM response to be both enhanced and red-shifted. Also, it is apparent to observe a flat region shown in Fig. 4.5(b), which results from the ω_2-related (rather than ω_1-related) local-field enhancement factor $|\bar{\varepsilon}(\omega_2)/\varepsilon(z, \omega_2)|^2$ for the whole graded film. The reason is that the factor $|\bar{\varepsilon}(\omega_2)/\varepsilon(z, \omega_2)|^2$ is approximately constant as ω_2 is small. This is further due to the above-mentioned fact that the effective linear dielectric constant of the graded film of interest is dominated by the layer with the smallest dielectric constant at small ω_2. In other words, as ω_2 is small, the corresponding effective linear dielectric constant is approximately the same as the smallest layer dielectric constant at small ω_2. Thus, the ω_2-related enhancement factor is almost constant for small ω_2. In this connection, a flat region should be displayed. In a word, although the enhancement of the effective NDFM susceptibility is often accompanied with the appearance of the optical absorption, the FOM is still possible to be quite attractive due to the presence of the gradation of the metallic film, see Fig. 4.5(c).

Figure 4.6 shows the results of model calculations in which a gradation profile of the relaxation rate of the form in Eq. (4.23) is also included. It is found that the relaxation-rate gradation plays an important role in the NDFM response as $\omega_2 \to \omega_p(0)$ or $\omega_2 \sim \omega_1[= 0.6\omega_p(0)]$. As a matter of fact, the enhancement in the NDFM response arises from the enhancement in the local field. In addition, by taking into account the relaxation-rate gradation, the FOM is also caused to increase at $\omega_2 \to \omega_p(0)$.

Since the present results do not depend crucially on the particular form of the dielectric function and the gradation profile and the point is that one needs a sufficiently large gradient, it is expected that an enhancement in NDFM responses will also be found in composi-

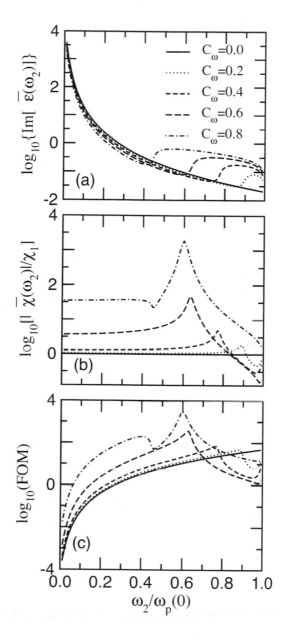

Figure 4.5. (a) The linear optical absorption $\mathrm{Im}[\bar{\varepsilon}(\omega_2)]$, (b) the enhancement of the NDFM susceptibility $|\bar{\chi}(\omega_2)|/\chi_1$, and (c) the FOM $\equiv |\bar{\chi}(\omega_2)|/\{\chi_1 \mathrm{Im}[\bar{\varepsilon}(\omega_2)]\}$ versus the normalized incident angular frequency $\omega_2/\omega_p(0)$ for dielectric function gradation profile $\varepsilon(z,\omega) = 1 - \omega_p^2(z)/[\omega(\omega+i\gamma(z))]$ with various plasma-frequency gradation profile $\omega_p(z) = \omega_p(0)(1 - C_\omega \cdot z)$ [Eq. (4.22)] and relaxation-rate gradation profile $\gamma(z) = \gamma(\infty) + C_\gamma/z$ [Eq. (4.23)], for various C_ω, at $\gamma(\infty) = 0.02\omega_p(0)$ and $\omega_1 = 0.6\omega_p(0)$. Parameter: $C_\gamma = 0.0$. After Ref. [88].

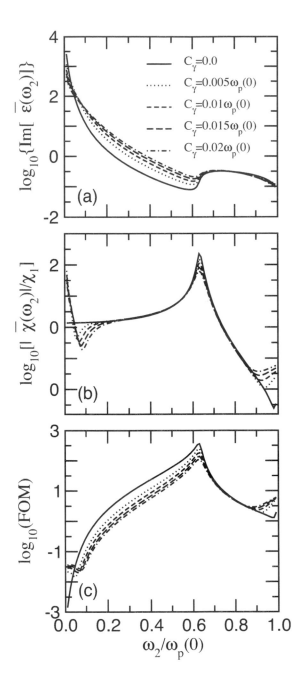

Figure 4.6. Same as Fig. 4.5, but for various C_γ. Parameter: $C_\omega = 0.6$. After Ref. [88].

tionally graded metal-dielectric composite films in which the fraction of metal component varies perpendicular to the film.

Finally, let us compare the NDFM susceptibility (Figs. 4.5-4.6) with the degenerate four-wave mixing (DFM) susceptibility (Figs. 4.1-4.2). We find that the resonant plasmon band in the linear optical absorption appears to be narrower for the NDFM susceptibility than for the DFM. Similarly, the plasmon band in the enhancement of the nonlinearity becomes much narrower for the NDFM susceptibility than for the DFM. In this connection, for the NDFM susceptibility, the plasmon band shown in the FOM is caused to be narrower, too. Such phenomena are caused to occur by the local-field effect. In addition, for both the NDFM and DFM susceptibilities, increasing C_γ causes the nonlinearity enhancement and FOM to be increased at the low probe frequencies which are generally smaller than ω_c [for the NDFM susceptibility $\omega_c \approx 0.05\omega_p(0)$; for the DFM susceptibility $\omega_c \approx 0.2\omega_p(0)(> 0.05\omega_p(0))$]. Also, owing to the local-field effect, the strength of the increment becomes lower for the NDFM susceptibility than for the DFM.

In the degenerate four-wave mixing setup, there are two pump fields E_f and E_b, and two probe fields E_p and E_s. Generally, the two pump fields are propagating in the opposite direction, i.e., the wave vectors satisfy $k_f + k_b = 0$, while the two probe fields are also opposite, i.e., $k_p + k_s = 0$. Thus the four wave vectors form a parallelogram. In this case, $\omega_p = \omega_s$ too, and we have calculated the NDFM susceptibility. However, we can also consider a more general case. Namely, the four wave vectors form a trapezoid. In this configuration, k_f and k_b, being along the two non-parallel sides, are not exactly opposite to each other, although $\omega_f = \omega_b$. In this case, k_p and k_s are parallel (opposite) to each other, but ω_p is not equal to ω_s. In addition, it is also worth calculating the NDFM susceptibility for this trapezoid case.

To sum up, we have investigated the surface plasmon resonant effect on the linear optical absorption, the enhancement in the NDFM response, and the FOM of graded metallic films. It is found that the presence of gradation in metallic films can yield a broad resonant plasmon band in the optical region, resulting in a large enhancement of the NDFM response and hence a large FOM.

2.   Multilayer Metallic Films

2.1.   Third-Order Nonlinearity

To discuss the multilayer effect on the effective nonlinear optical response, let us first start from a general case, i.e., graded metallic film. In detail, we consider a graded metallic film with width L, and its gradation is in the direction perpendicular to the film. As a matter of fact, for graded films, the formalism has been derived in Section 1.1.. For completeness, below we shall do a brief review.

Inside the graded film, the local constitutive relation between the displacement $\mathbf{D}$ and the electric field $\mathbf{E}$ is given by

$$\mathbf{D}(z, \omega) = \varepsilon(z, \omega)\mathbf{E}(z, \omega) + \chi(z, \omega)|\mathbf{E}(z, \omega)|^2\mathbf{E}(z, \omega), \qquad (4.24)$$

where $\varepsilon(z, \omega)$ and $\chi(z, \omega)$ stand for the linear dielectric constant and third-order nonlinear susceptibility, respectively, and both are gradation profiles as a function of position z.

Again, the weak nonlinearity condition is assumed to be satisfied. In the quasi-static approximation, the whole graded film can be regarded as an effective homogeneous one with effective linear dielectric constant $\bar{\varepsilon}(\omega)$ and effective third-order nonlinear susceptibility $\bar{\chi}(\omega)$. Both $\bar{\varepsilon}(\omega)$ and $\bar{\chi}(\omega)$ are defined as [66]

$$\langle \mathbf{D} \rangle = \bar{\varepsilon}(\omega)\mathbf{E}_0 + \bar{\chi}(\omega)|\mathbf{E}_0|^2\mathbf{E}_0, \tag{4.25}$$

where $\langle \cdots \rangle$ denotes the spatial average, and $\mathbf{E}_0 = E_0\hat{e}_z$ is the applied field along z axis.

Then, we adopt the following graded Drude dielectric profile

$$\varepsilon(z,\omega) = 1 - \frac{\omega_p^2(z)}{\omega(\omega+i\gamma)}, \tag{4.26}$$

where $0 \leq z \leq L$, and γ stands for the damping coefficient in the corresponding bulk material. The general form in Eq. (4.26) allows for the possibility of a gradation profile in the plasma frequency $\omega_p(z)$ [e.g., Eq. (4.4)].

In view of the $z-$dependent profile, let us use the equivalent capacitance for capacitors in series to evaluate the effective perpendicular linear response for a given frequency $\bar{\varepsilon}(\omega)$, [14]

$$\frac{1}{\bar{\varepsilon}(\omega)} = \frac{1}{L}\int_0^L \frac{dz}{\varepsilon(z,\omega)}. \tag{4.27}$$

Next, we take one step forward to write the effective nonlinear response $\bar{\chi}(\omega)$ as an integral over the film, [14]

$$\bar{\chi}(\omega) = \frac{1}{L}\int_0^L dz \chi(z,\omega)\left|\frac{\bar{\varepsilon}(\omega)}{\varepsilon(z,\omega)}\right|^2 \left(\frac{\bar{\varepsilon}(\omega)}{\varepsilon(z,\omega)}\right)^2 \tag{4.28}$$

where $\chi(z,\omega)$ denotes the local third-order nonlinear susceptibility for a given frequency. It is worth noting that the real $\bar{\chi}(\omega)$ should involve an integral over x, y, and z of the local $\chi(x,y,z,\omega)$ multiplied by terms involving $\varepsilon(x,y,z,\omega)$. Thus, Eq. (4.28) offers an approximate $\bar{\chi}(\omega)$, as expected.

To investigate the multilayer effect, we shall use some finite difference approximation of the graded Drude profile [Eq. (4.26)] for a finite number of layers.

To mimic a multilayer system (Fig. 4.7), we divide the interval $[0,L]$ into N equally spaced sub-intervals, $[0,z_1], (z_1,z_2], \cdots, (z_{N-1},z_N]$. Then we adopt the midpoint value of $\omega_p(z)$ for each sub-interval as the plasma frequency of that sublayer. In this way, we calculate the effective dielectric constant, the effective third-order nonlinear susceptibility, as well as the figure of merit for each N. It is worth noting that for $N \to \infty$ (e.g., $N = 1024$) the graded film results [14] recover in this limit.

In what follows, we shall do some numerical calculations. We assume that the metal layers within the film have the same real and positive frequency-independent third-order nonlinear susceptibility $\chi(z,\omega) = \chi_1$, and do not have a gradation profile. In doing so, we could focus on the enhancement of the optical nonlinearity. Without loss of generality, the film width L is taken to be unity.

For numerical calculations, we take as a model plasma-frequency gradation profile

$$\omega_p(z) = \omega_p(0)(1 - C_\omega \cdot z), \tag{4.29}$$

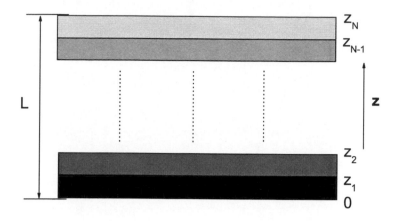

Figure 4.7. Schematic graph showing multilayer metallic films.

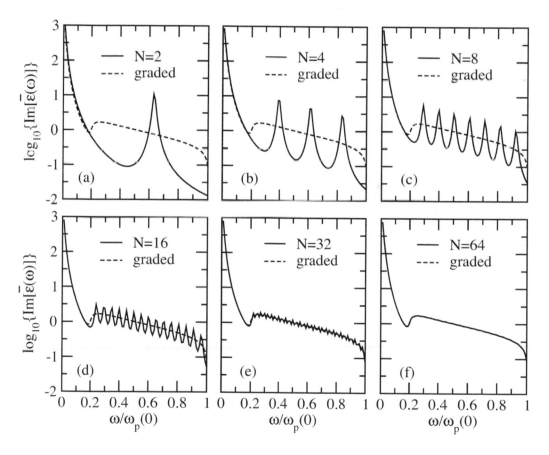

Figure 4.8. The linear optical absorption $\mathrm{Im}[\bar{\varepsilon}(\omega)]$ versus the normalized incident angular frequency $\omega/\omega_p(0)$ for dielectric function gradation profile $\varepsilon(z, \omega) = 1 - \omega_p^2(z)/[\omega(\omega + i\gamma)]$ with various plasma-frequency gradation profile $\omega_p(z) = \omega_p(0)(1 - C_\omega \cdot z)$. Parameters: $\gamma = 0.02\omega_p(0)$, $C_\omega = 0.8$, $L = 1$, and $\chi_1 = 1$. After Ref. [89].

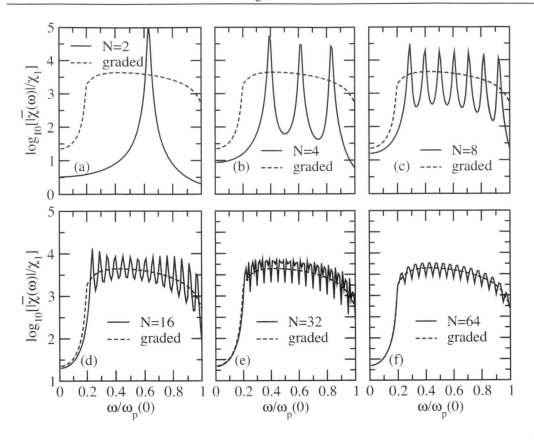

Figure 4.9. Same as Fig. 4.8, but for the enhancement of the third-order optical nonlinearity $|\bar{\chi}(\omega)|/\chi_1$. After Ref. [89].

where C_ω is a constant (gradient) tuning the profile.

Figures 4.8-4.10 respectively display the optical absorption $\sim \mathrm{Im}[\bar{\varepsilon}(\omega)]$, the modulus of the effective third-order optical nonlinearity enhancement $|\bar{\chi}(\omega)|/\chi_1$, as well as the FOM $|\bar{\chi}(\omega)|/\{\chi_1\mathrm{Im}[\bar{\varepsilon}(\omega)]\}$ as a function of frequency $\omega/\omega_p(0)$. Here $\mathrm{Im}[\cdots]$ means the imaginary part of $\cdots$. In each panel of Figs. 4.8-4.10, the corresponding graded film results are shown as well.

It is evident from Figs. 4.8-4.10 that for a few layers, say $N = 2, 4, 8$ [(a) - (c)], the optical absorption spectrum and the enhancement of optical nonlinearity consist mainly of sharp peaks. However, the strong optical absorption and the large fluctuation of the nonlinear optical enhancement near these sharp peaks render the FOM too small to be useful. When the number of layers becomes large [(d) - (f)], the sharp peaks accumulate to a broad band while the fluctuation has been reduced significantly. In this limit, the broad continuous absorption band emerges, and a large FOM persists.

In a word, all figures 4.8-4.10 show a gradual transition from sharp peaks to a broad continuous band as the number of layers increases. This also gives an explanation of the intriguing findings in Chapter 1.1..

In this section, we have investigated the effective nonlinear optical response of metallic films as the number of layers inside the film increases until the graded film results recover.

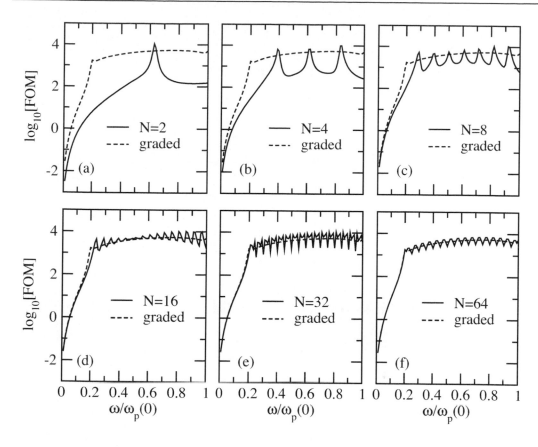

Figure 4.10. Same as Fig. 4.8, but for the FOM $\equiv |\bar{\chi}(\omega)|/\{\chi_1 \mathrm{Im}[\bar{\varepsilon}(\omega)]\}$. After Ref. [89].

This is of practical value since in practice it is more convenient to fabricate multilayer metallic films than graded films.

Also, it is of interest to extend the present consideration to composites in which graded spherical particles are embedded in a host medium to account for mutual interactions among graded particles. Similar considerations can be extended to other nonlinear optical properties like the second-harmonic generation [16] and strong nonlinearity [90].

To sum up, we have investigated the multilayer effect on the effective nonlinear optical response of metallic films. As the number of layers inside the metallic films increases, a gradual transition has been shown from sharp peaks to a broad continuous band until the graded film results recover.

Chapter 5

Graded Composites

Graded materials are those whose properties vary gradually as a function of position. This gradation may occur naturally or may be a product of manufacturing processes. It was reported [4, 5, 14, 16] that graded (inhomogeneous) materials can show stronger nonlinear optical responses than the corresponding homogeneous ones. Also, it is desirable and of interest to use dielectric-coated metallic nanoparticles with varying shell thickness to form a dielectric constant gradient [42].

Graded materials have many applications since the gradation profile provides an additional means of controlling the physical properties. In other words, graded materials can have quite different physical properties from homogeneous materials [12, 16, 43, 63, 91, 92, 93]. To achieve desired optical responses, the technique of gradation is expected to be useful for a variety of optical materials, e.g., metal-dielectric composites and negative-refractive-index metamaterials [22, 94]. In particular, graded thin films often possess different optical properties, when compared with a film processing the same properties at different locations along the growth direction. For example, a large enhancement in nonlinear optical responses was found in a composite of alternating, sub-wavelength-thick layers of titanium dioxide and conjugated polymer [4], which can be regarded as a graded material. It has also been observed that compositionally graded barium strontium titanate thin films have better electric properties than a single-layer barium strontium titanate film with the same composition [59].

1.   Graded Metal-Dielectric Composite Films

Chemical deposition techniques [Fig. 5.1(a)] as well as diffusing techniques [Fig. 5.1(b)] can be used to produce inhomogeneous composite films or interfaces. Metal-dielectric composites have received much attention due to the potential application of their linear and nonlinear optical properties [4, 55, 66, 67, 68, 69, 70, 71, 73, 95, 96, 97, 98, 99, 100]. Crucial elements for control of these properties are the micro-structure of the composite, particle shape, and the properties of the constituents. For anisotropically shaped metallic nanoparticles, the resonant plasmon bands split up for orientations along major and minor axes. Furthermore, in case of a large size aspect ratio, the plasmon bands may shift into the infrared, thus enabling the use of metal nanostructures in telecommunication applications in this wavelength range. Compared with spherically shaped particles, anisotropically shaped metallic particles can show reduced plasmon relaxation times [101] as well as enhanced nonlinear responses [102], and may thus be used as building blocks in a variety of optical devices. Some techniques have been developed to fabricate rod-shaped metallic nanoparticles by using lithographic means [103] or anisotropic growth. Recently, one has demonstrated that mega-electron-volt ion irradiation can also be used to modify the shape of nanoparticles [104]. This ion-beam-induced anisotropic deformation effect is known to occur not only for a broad range of amorphous materials [105], but also for crystalline materials including metals [100]. That is, prolate spheroidal metallic particles in a dielectric host can readily be formed by irradiation of mega-electron-volt ions.

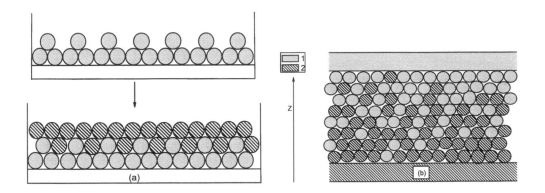

Figure 5.1. Schematic graph showing (a) the chemical deposition technique to form a graded composite film, and (b) the diffusing technique to form a graded interface (composite film) between two bulk composites. After Ref. [106].

The optical property of thin films [14, 16, 58, 61, 107] is often different from that of the corresponding bulk material. Recently, a metal-dielectric composite (Au:BaTiO$_3$) film was investigated, and a large FOM (figure of merit) was observed in Ref. [108]. If there were a larger third-order nonlinear susceptibility and/or a smaller linear optical absorption under certain conditions, the corresponding FOM should be relatively larger, thus being attractive. However, the surface-plasmon resonant nonlinearity enhancement often occurs concomitantly with a strong absorption, and unfortunately this behavior renders the FOM of the resonant enhancement peak to be too small to be useful. To circumvent this problem, we shall consider a kind of graded metal-dielectric composite film, in which a dielectric (or metallic) component is introduced as anisotropically shaped particles embedded in a metallic (or dielectric) component with a compositional and/or shape-dependent gradation profile.

For nonlinear dielectric phenomena in Kerr composites, the nonlinear alternating current responses have been investigated [109, 110, 111, 112]. For nonlinear optical effects other than the Kerr effect, Hui *et al.* [75] derived general expressions for the effective susceptibility for the second-harmonic generation (SHG) in a binary composite of random dielectrics. Recently, these authors have also studied the thickness dependence of the effective SHG susceptibility in films of random dielectrics [76] and the enhancement in the SHG in dilute composites of coated small particles [77].

To achieve an enhanced and controllable SHG in functional optical materials is still a challenging task (see Refs. [16, 79, 113] and references therein). In the following, we shall investigate the effects of gradation and/or particle shape on the SHG response in graded metal-dielectric films. In these films, the volume fraction of the anisotropically shaped metallic particles varies in the direction perpendicular to the film, i.e., along the growth direction of the film.

1.1. Third-Order Nonlinearity

Let us consider a metal-dielectric composite film with a variation of volume fraction of anisotropic particles along z axis perpendicular to the film (Fig. 5.2). In this case, the local constitutive relation between the displacement field $\mathbf{D}(\mathbf{r}, \omega)$ and electric field $\mathbf{E}(\mathbf{r}, \omega)$ is given by

$$\mathbf{D}(\mathbf{r}, \omega) = \varepsilon(\mathbf{r}, \omega)\mathbf{E}(\mathbf{r}, \omega) + \chi(\mathbf{r}, \omega)|\mathbf{E}(\mathbf{r}, \omega)|^2 \mathbf{E}(\mathbf{r}, \omega), \qquad (5.1)$$

where $\varepsilon(\mathbf{r}, \omega)$ and $\chi(\mathbf{r}, \omega)$ are the linear dielectric constant and third-order nonlinear susceptibility of a layer inside the graded film, respectively. Here both $\varepsilon(\mathbf{r}, \omega)$ and $\chi(\mathbf{r}, \omega)$ are the gradation profiles as a function of position $\mathbf{r}$ and field frequency ω.

Here the weak nonlinearity condition is assumed to be satisfied. That is, the contribution of the second term [nonlinear part $\chi(\mathbf{r}, \omega)|\mathbf{E}(\mathbf{r}, \omega)|^2$] in the right-hand side of Eq. (5.1) is much less than that of the first term [linear part $\varepsilon(\mathbf{r}, \omega)$] [66]. Next, we turn to the quasi-static approximation, under which the whole graded structure can be regarded as an effective homogeneous one with effective linear dielectric constant $\bar{\varepsilon}(\omega)$ and effective third-order nonlinear susceptibility $\bar{\chi}(\omega)$. In this connection, $\bar{\varepsilon}(\omega)$ and $\bar{\chi}(\omega)$ are defined as [66]

$$\mathbf{D}_0 = \bar{\varepsilon}(\omega)\mathbf{E}_0 + \bar{\chi}(\omega)|\mathbf{E}_0|^2 \mathbf{E}_0, \qquad (5.2)$$

where $\mathbf{D}_0$ and $\mathbf{E}_0 (= E_0 \hat{\mathbf{e}}_z)$ are respectively the volume-averaged displacement field and electric field within the whole graded composite film.

For calculating the nonlinear optical response, we shall apply a two-step solution. In Step A, we first derive the responses of a layer inside the graded film, $\bar{\varepsilon}(z, \omega)$ and $\bar{\chi}(z, \omega)$. In Step B, the overall responses of the graded film, $\bar{\varepsilon}(\omega)$ and $\bar{\chi}(\omega)$, are derived. In the two-step solution, it should be remarked that the local field inside the spheroidal particles is uniform, and the effective nonlinear response of a layer is therefore exact within the Maxwell-Garnett theory. When we have a nonlinear host, we shall have to invoke the decoupling approximation [114]. It is worth noting that Step B is exact, see Eqs. (5.16)-(5.18) below. In a word, in Step A, we homogenize the composite film along xy plane while in Step B, we further homogenize the graded film along z axis.

A. Responses of a layer inside the graded film: $\bar{\varepsilon}(z, \omega)$ *and* $\bar{\chi}(z, \omega)$

It is not possible to calculate $\bar{\varepsilon}(z, \omega)$ exactly in terms of the compositional and/or shape-dependent gradation profile. Nevertheless, to obtain an estimate of $\bar{\varepsilon}(z, \omega)$, we can take a small volume element inside the layer, at a position z. Further, this small volume element can be seen as a composite where the locations of the inclusion particles are random in the host medium. This, however, is a highly directional distribution since the long or short axis of prolate or oblate particles is parallel to the gradient along z axis. Accordingly, the volume fraction of the inclusion is $p(z)$ for the dielectric or $1 - p(z)$ for the metal. In this regard, the well-known Maxwell-Garnett approximation holds very well for computing $\bar{\varepsilon}(z, \omega)$ [as shown in Eqs. (5.3) and (5.4) below]. In detail, for the dielectric particles embedded in the metallic component in a layer, $\bar{\varepsilon}(z, \omega)$ can be given by the first-kind of Maxwell-Garnett approximation (MGA1) [23, 24, 115]

$$\frac{\bar{\varepsilon}(z, \omega) - \varepsilon_1(\omega)}{L_z^{(2)} \bar{\varepsilon}(z, \omega) + (1 - L_z^{(2)})\varepsilon_1(\omega)} = p(z) \frac{\varepsilon_2 - \varepsilon_1(\omega)}{L_z^{(2)} \varepsilon_2 + (1 - L_z^{(2)})\varepsilon_1(\omega)}, \qquad (5.3)$$

where $L_z^{(2)}$ is the depolarization factor of the dielectric particles along $z-$axis, and satisfies a sum rule $L_z^{(2)} + 2L_x^{(2)} = 1$. Here $L_x^{(2)}$ is the depolarization factor of the dielectric particles along $x(y)-$axis. For more details on the depolarization factors of particles, we refer the read to Appendix B. In Eq. (5.3), ε_2 (or $\varepsilon_1(\omega)$) stands for the dielectric constant of the dielectric (or metallic) particles, and $p(z)$ denotes the volume fraction of the dielectric particles in each layer which is thus a compositional gradation profile as a function of position z. Alternatively, for the metallic particles embedded in the dielectric host, the second kind of Maxwell-Garnett approximation (MGA2) can be used to determine $\bar{\varepsilon}(z, \omega)$, such that [23, 24, 115]

$$\frac{\bar{\varepsilon}(z, \omega) - \varepsilon_2}{L_z^{(1)} \bar{\varepsilon}(z, \omega) + (1 - L_z^{(1)})\varepsilon_2} = (1 - p(z)) \frac{\varepsilon_1(\omega) - \varepsilon_2}{L_z^{(1)} \varepsilon_1(\omega) + (1 - L_z^{(1)})\varepsilon_2}, \qquad (5.4)$$

where $L_z^{(1)}$ is the depolarization factor of the metallic particles along $z-$axis. Similarly, there exists $L_z^{(1)} + 2L_x^{(1)} = 1$ where $L_x^{(1)}$ is the depolarization factor of the metallic particles along $x(y)-$axis. It is worth noting that $L_z < 1/3, = 1/3$ and $> 1/3$ indicates the fact that the metallic (or dielectric) particles exist in the form of prolate spheroid, sphere and oblate spheroid, respectively. In Eqs. (5.3) and (5.4), the dielectric constant of the metal $\varepsilon_1(\omega)$ is given by the known Drude expression

$$\varepsilon_1(\omega) = 1 - \frac{\omega_p^2}{\omega(\omega + i\gamma)}, \qquad (5.5)$$

where ω_p denotes the bulk plasmon frequency, and γ the collision frequency. In case of $\varepsilon_1(\omega) > \varepsilon_2$, the MGA1 always gives an upper bound while the MGA2 a lower bound, and vice versa. The exact result must lie between the two bounds [116]. For both the MGA1 and MGA2, the particles under discussion are randomly embedded but their orientations are all along z axis (i.e., perpendicular to the film). The reason is that in experiments the prolate spheroidal metallic particles can be highly oriented along the direction of irradiated ions [100]. For completeness, we shall also numerically calculate the case of oblate spheroids.

Then, we calculate the effective nonlinear response for a layer at position z, $\bar{\chi}(z, \omega)$ [66],

$$\bar{\chi}(z, \omega)\mathbf{E}(z)^4 = (1 - p(z))\chi_1 \langle |\mathbf{E}_1(z)|^2 \mathbf{E}_1(z)^2 \rangle + p(z)\chi_2 \langle |\mathbf{E}_2(z)|^2 \mathbf{E}_2(z)^2 \rangle, \qquad (5.6)$$

where χ_1 and χ_2 are respectively the (intrinsic) third-order nonlinear susceptibility of the metallic and dielectric components, $E_1(z)$ (or $E_2(z)$) represents the local electric field inside the metallic (or dielectric) component within a layer at position z, $E(z)$ denotes the volume-averaged electric field within the layer, and $\langle \cdots \rangle$ stands for the volume average of $\cdots$ within

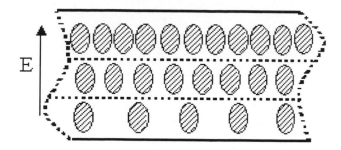

Figure 5.2. Schematic graph showing the geometry of a metal-dielectric composite film with a variation of volume fraction of anisotropic particles along z axis perpendicular to the film. The electric field E is parallel to the gradient (z axis), thus being perpendicular to the film. After Ref. [117].

the layer. In order to estimate $\bar{\chi}(z,\omega)$, due to the existence of nonlinear host we have to invoke the decoupling approximation [114]

$$\langle |\mathbf{E}_i(z)|^2 \mathbf{E}_i(z)^2 \rangle = \langle |\mathbf{E}_i(z)|^2 \rangle \langle \mathbf{E}_i(z)^2 \rangle, \ i = 1, 2. \tag{5.7}$$

For the sake of consistency, the local field averages $\langle |\mathbf{E}_i(z)|^2 \rangle$ and $\langle \mathbf{E}_i(z)^2 \rangle$ should be determined by using the Maxwell-Garnett technique [118]. For the MGA1, we obtain the local field averages such that

$$\langle \mathbf{E}_2(z)^2 \rangle = \frac{(L_z^{(2)})^{-2} \varepsilon_1(\omega)^2}{[(1-p(z))\varepsilon_2 + ((L_z^{(2)})^{-1} - (1-p(z)))\varepsilon_1(\omega)]^2} \mathbf{E}(z)^2, \tag{5.8}$$

$$\langle \mathbf{E}_1(z)^2 \rangle = \theta \left(1 - \frac{p(z)(L_z^{(2)})^{-1}[(L_z^{(2)})^{-1}\varepsilon_1(\omega)^2 - (1-p(z))(\varepsilon_2 - \varepsilon_1(\omega))^2]}{[(1-p(z))\varepsilon_2 + ((L_z^{(2)})^{-1} - (1-p(z)))\varepsilon_1(\omega)]^2} \right) \mathbf{E}(z)^2, \tag{5.9}$$

$$\langle |\mathbf{E}_2(z)|^2 \rangle = \frac{(L_z^{(2)})^{-2}|\varepsilon_1(\omega)|^2}{|(1-p(z))\varepsilon_2 + ((L_z^{(2)})^{-1} - (1-p(z)))\varepsilon_1(\omega)|^2} \mathbf{E}(z)^2, \tag{5.10}$$

$$\langle |\mathbf{E}_1(z)|^2 \rangle = \theta \left(1 - \frac{p(z)(L_z^{(2)})^{-1}[(L_z^{(2)})^{-1}|\varepsilon_1(\omega)|^2 - (1-p(z))|\varepsilon_2 - \varepsilon_1(\omega)|^2]}{|(1-p(z))\varepsilon_2 + ((L_z^{(2)})^{-1} - (1-p(z)))\varepsilon_1(\omega)|^2} \right) \mathbf{E}(z)^2, \tag{5.11}$$

with $\theta = 1/(1-p(z))$. Similarly, for the MGA2, the local field averages are given by

$$\langle \mathbf{E}_1(z)^2 \rangle = \frac{(L_z^{(1)})^{-2}\varepsilon_2^2}{[p(z)\varepsilon_1(\omega) + ((L_z^{(1)})^{-1} - p(z))\varepsilon_2]^2} \mathbf{E}(z)^2, \tag{5.12}$$

$$\langle \mathbf{E}_2(z)^2 \rangle = \theta' \left(1 - \frac{(1-p(z))(L_z^{(1)})^{-1}[(L_z^{(1)})^{-1}\varepsilon_2^2 - p(z)(\varepsilon_1(\omega) - \varepsilon_2)^2]}{[p(z)\varepsilon_1(\omega) + ((L_z^{(1)})^{-1} - p(z))\varepsilon_2]^2} \right) \mathbf{E}(z)^2, \tag{5.13}$$

$$\langle |\mathbf{E}_1(z)|^2 \rangle = \frac{(L_z^{(1)})^{-2}|\varepsilon_2|^2}{|p(z)\varepsilon_1(\omega) + ((L_z^{(1)})^{-1} - p(z))\varepsilon_2|^2} \mathbf{E}(z)^2, \tag{5.14}$$

$$\langle |\mathbf{E}_2(z)|^2 \rangle = \theta' \left(1 - \frac{(1-p(z))(L_z^{(1)})^{-1}[(L_z^{(1)})^{-1}|\varepsilon_2|^2 - p(z)|\varepsilon_1(\omega) - \varepsilon_2|^2]}{|p(z)\varepsilon_1(\omega) + ((L_z^{(1)})^{-1} - p(z))\varepsilon_2|^2} \right) \mathbf{E}(z)^2, \tag{5.15}$$

with $\theta' = 1/p(z)$.

B. Overall responses of the graded film: $\bar{\varepsilon}(\omega)$ and $\bar{\chi}(\omega)$

Owing to the simple graded structure (Fig. 5.2), we can use the equivalent capacitance of series combination to calculate the linear response (i.e., optical absorption for the graded

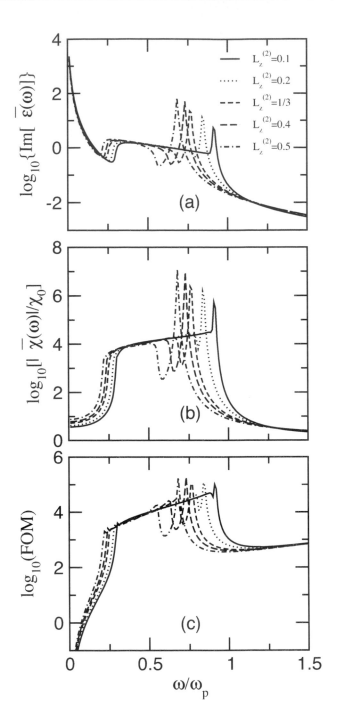

Figure 5.3. *Results for the MGA1 [Eq. (5.3)].* (a) Linear optical absorption $\mathrm{Im}[\bar{\varepsilon}(\omega)]$, (b) enhancement of the third-order optical nonlinearity $|\bar{\chi}(\omega)|/\chi_0$, and (c) FOM $\equiv$ $|\bar{\chi}(\omega)|/\{\chi_0\mathrm{Im}[\bar{\varepsilon}(\omega)]\}$ versus the normalized incident angular frequency ω/ω_p for layer dielectric profile $p(z) = az^m$, for different $L_z^{(2)}$. Parameters: $a = 0.8$, $m = 1$, $\gamma/\omega_p = 0.01$, and $\varepsilon_2 = (3/2)^2$. After Ref. [117].

film), $\bar{\varepsilon}(\omega)$,

$$\frac{1}{\bar{\varepsilon}(\omega)} = \frac{1}{W} \int_0^W \frac{dz}{\bar{\varepsilon}(z,\omega)}, \tag{5.16}$$

where W is the thickness of the film.

By virtue of the continuity of electric displacement, there is a relation

$$\bar{\varepsilon}(z,\omega)E(z) = \bar{\varepsilon}(\omega)E_0. \tag{5.17}$$

Then, we take one step forward to obtain the effective third-order nonlinear susceptibility $\bar{\chi}(\omega)$ as an integral over the graded film,

$$\bar{\chi}(\omega) = \frac{1}{W} \int_0^W dz \bar{\chi}(z,\omega) \left| \frac{\bar{\varepsilon}(\omega)}{\bar{\varepsilon}(z,\omega)} \right|^2 \left(\frac{\bar{\varepsilon}(\omega)}{\bar{\varepsilon}(z,\omega)} \right)^2. \tag{5.18}$$

In what follows, we shall do some numerical calculations. Set both χ_1 and χ_2 to be a real and positive frequency-independent constant χ_0, so that we could emphasize the enhancement of the optical nonlinearity. Without loss of generality, the film thickness W is taken to be unity. For the model calculation, we shall use the gradation profile

$$p(z) = az^m, \tag{5.19}$$

where a and m are constants tuning the profile.

Figure 5.3 shows the results obtained from the MGA1 [Eq. (5.3)]. In this figure, we display (a) the optical absorption $\sim \mathrm{Im}[\bar{\varepsilon}(\omega)]$, (b) the modulus of the effective third-order optical nonlinearity enhancement $|\bar{\chi}(\omega)|/\chi_0$, and (c) the FOM $|\bar{\chi}(\omega)|/\{\chi_0\mathrm{Im}[\bar{\varepsilon}(\omega)]\}$ as a function of the incident angular frequency ω, for different $L_z^{(2)}$. Here $\mathrm{Im}[\cdots]$ means the imaginary part of $\cdots$. When the layer gradation profile $p(z) = az^m$ is taken into account, a broad resonant plasmon band is observed for any $L_z^{(2)}$ of interest. In other words, the broad band is caused to appear by the effect of the positional dependence of the dielectric or metallic component. This conclusion may be made by comparing the curves in Fig. 5.3 with those of $n = 0$ (corresponding to the case where the effects of gradation as well as non-spherical shape are excluded) in Fig. 5.8. Moreover, we find that *decreasing* $L_z^{(2)}$ makes the resonant bands in both optical nonlinearity and optical absorption broader. Although the enhancement of the effective third-order optical nonlinearity is often accompanied with the appearance of the optical absorption, the FOM is still possible to be very attractive [see Fig. 5.3(c)] due to the positional dependence of the dielectric or metallic components. In particular, the particle shape can also be used to enhance the FOM significantly. It is worth noting that there is a prominent surface plasmon resonant peak which appears at somewhat higher frequencies in addition to the surface plasmon band at lower frequencies.

Figure 5.4 displays the results which were obtained from the MGA2 [Eq. (5.4)]. For this figure, we see the parallel shaped metallic particles are randomly embedded in the dielectric host in each layer. In contrast, the surface plasmon resonant peak is found to locate at lower frequencies in addition to the surface plasmon band which locates at higher frequencies. Also, it is shown that the broad plasmon bands in optical nonlinearity and absorption is caused to appear by the effect of gradation, when comparing the curves in Fig. 5.4 with those of $n = 0$ in Fig. 5.9. Note the latter corresponds to the case without the

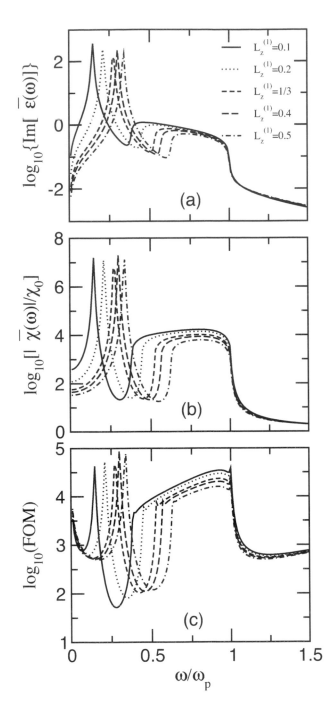

Figure 5.4. *Results for the MGA2 [Eq. (5.4)].* (a) Linear optical absorption $\text{Im}[\bar{\varepsilon}(\omega)]$, (b) enhancement of the third-order optical nonlinearity $|\bar{\chi}(\omega)|/\chi_0$, and (c) FOM $\equiv$ $|\bar{\chi}(\omega)|/\{\chi_0 \text{Im}[\bar{\varepsilon}(\omega)]\}$ versus the normalized incident angular frequency ω/ω_p for layer dielectric profile $p(z) = az^m$, for different $L_z^{(1)}$. Parameters: $a = 0.8$, $m = 1$, $\gamma/\omega_p = 0.01$, and $\varepsilon_2 = (3/2)^2$. After Ref. [117].

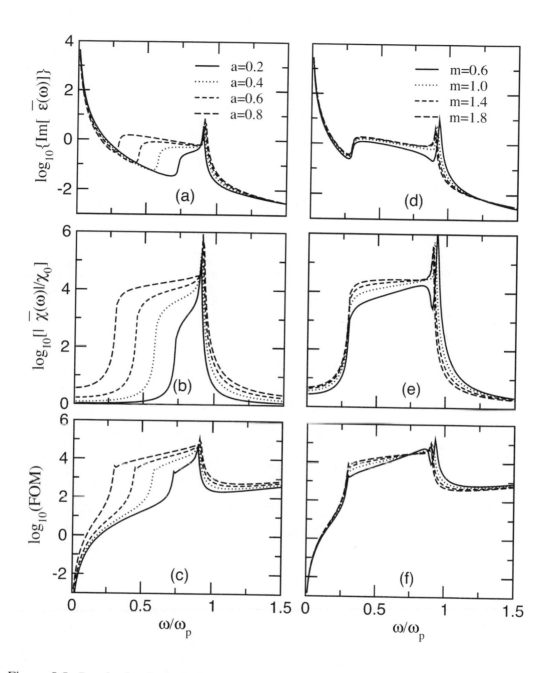

Figure 5.5. *Results for the MGA1 [Eq. (5.3)].* (a) and (d) $\text{Im}[\bar{\varepsilon}(\omega)]$, (b) and (e) $|\bar{\chi}(\omega)|/\chi_0$, and (c) and (f) $\text{FOM} \equiv |\bar{\chi}(\omega)|/\{\chi_0 \text{Im}[\bar{\varepsilon}(\omega)]\}$ versus ω/ω_p (a)-(c) for different a at $m = 1.0$, and (d)-(f) for different m at $a = 0.8$. Parameters: $L_z = 0.1$, $\gamma/\omega_p = 0.01$, and $\varepsilon_2 = (3/2)^2$. After Ref. [117].

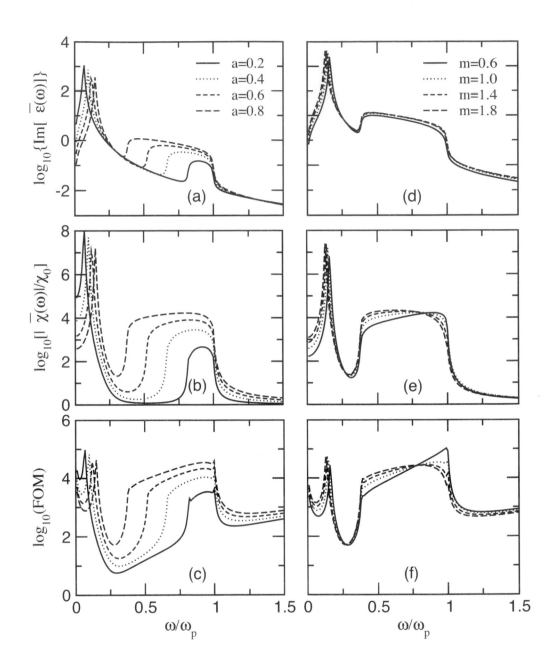

Figure 5.6. *Results for the MGA2 [Eq. (5.4)].* (a) and (d) $\text{Im}[\bar{\varepsilon}(\omega)]$, (b) and (e) $|\bar{\chi}(\omega)|/\chi_0$, and (c) and (f) $\text{FOM} \equiv |\bar{\chi}(\omega)|/\{\chi_0\text{Im}[\bar{\varepsilon}(\omega)]\}$ versus ω/ω_p (a)-(c) for different a at $m = 1.0$, and (d)-(f) for different m at $a = 0.8$. Parameters: $L_z = 0.1$, $\gamma/\omega_p = 0.01$, and $\varepsilon_2 = (3/2)^2$. After Ref. [117].

effects of gradation and non-spherical shape. Decreasing $L_z^{(1)}$ causes the plasmon bands to be broadened. This effect makes the FOM can be very attractive.

The MGA1 was applied to plot Fig. 5.5, in an attempt to discuss the effect of the gradation of the volume fraction of the dielectric by means of the gradation profile $p(z) = az^m$ for different (a-c) a and (d-f) m. In other words, we investigate a compositional gradation profile in the film, in which the dielectric particles possess a positional-dependent volume fraction. In detail, increasing a causes the resonant plasmon bands in optical nonlinearity and absorption to be broadened, see Fig. 5.5(a-b). Accordingly, in case of gradation, the FOM can be more attractive, see Fig. 5.5(c). Similarly, Fig. 5.5(d-f) displays the influence of m. It is apparent to see that the broad resonant plasmon bands in optical nonlinearity and absorption can be enhanced by *increasing m* [see Fig. 5.5(d-e)]. Finally, the FOM can become more attractive in the frequency range $0.3\omega_p < \omega < 0.7\omega_p$, as m increases, see Fig. 5.5(f).

Similar to Fig. 5.5, Fig. 5.6 is plotted via the MGA2. First, we discuss the effect of a. In both optical nonlinearity and absorption, the resonant plasmon bands are caused to be both enhanced and broadened by increasing a, as yielding a very attractive FOM in Fig. 5.6(c). On the other hand, the m effect on the optical nonlinearity and FOM plays a role too [see Fig. 5.6(e-f)], and accordingly both the optical nonlinearity and FOM can be enhanced accordingly. In addition, as a or m varies, the plasmon resonant peaks in Fig. 5.6 have the same red-shift (located at lower frequency) or blue-shift (located at higher frequency) behavior as those shown in Fig. 5.5 where the MGA1 was used instead.

During ion irradiation, the ion energy can be much larger at the top of the film than that at the bottom. Therefore, the particles can be much prolate at the top, but they are relatively spherical at the bottom. In other words, both $L_z^{(1)}$ and $L_z^{(2)}$ can be small at the top of the film, while increases to roughly $1/3$ at the bottom of the film. In this regard, we could introduce a gradation in the depolarization factor (Fig. 5.7) rather than in the volume fraction. Namely, in this case, $L_z(z)$ is a function of z (Fig. 5.7). For convenience, we keep the volume fraction to be constant [e.g., $p(z) = 0.85$] for each layer throughout the film, and take a physical profile $L_z(z) = (1/3)z^n$. In particular, as $n = 0$, we have $L_z(z) = 1/3$, i.e., the gradation in the depolarization factor and the non-spherical shape effect disappear. For different n, the corresponding results are shown in Figs. 5.8 and 5.9 for the MGA1 and MGA2, respectively. It is shown that the $L_z(z)$ profile does have a significant impact on the optical response, as expected. In Fig. 5.8(a), the plasmon peak shows a reduction as well as a blue-shift as n changes from zero (without gradation) to nonzero (with gradation), and accordingly the optical nonlinearity and hence the FOM is reduced. The difference between the results for different nonzero n (i.e., $n = 4, 8, 12$) is not very distinct (Fig. 5.8). Interestingly, the L_z gradation gives rise to an additional peak which appears at a lower frequency. For the MGA2 (Fig. 5.9), the surface plasmon resonant bands in optical absorption and nonlinearity are clearly visible for various $L_z(z)$ profiles, see Fig. 5.9(a-b). In the presence of gradation, i.e., n becomes nonzero, the prominent plasmon absorption peak at $n = 0$ has been broadened into a plasmon band, and an additional peak is induced to appear at lower frequency. Concomitantly, a plasmon band and a peak in optical nonlinearity are also caused to appear [Fig. 5.9(b)] and hence the FOM can be enhanced accordingly, see Fig. 5.9(c). On the other hand, we also find that the plasmon bands in optical absorption and nonlinearity can be further broadened (and enhanced) by adopting a wider gradation profile such as $L_z(z) = 0.5z^n$ (no figures shown here). For this sort of profile, there are prolate particles at the top, but oblate particles at the bottom of the film. It is possible to realize such oblate particles near the bottom of the film due to the reaction stress from the substance.

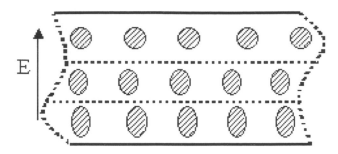

Figure 5.7. Schematic graph to show the geometry of a metal-dielectric composite film with a variation of depolarization factor of particles along z axis perpendicular to the film. The electric field E is parallel to the gradient (z axis), thus being perpendicular to the film. After Ref. [117].

Finally, in Figs. 5.4(a), 5.6(a) and 5.9(a), there are always a plasmon band plus an absorption peak as long as the gradation profile exists. Recently, an absorption peak plus a slim plasmon absorption band was indeed observed [100], when one investigated the optical extinction spectra for ensembles of core-shell colloids with Au cores and shells embedded in an index-matching fluid. But, after irradiation with 30 MeV (mega electron volt) Cu ions, a broadening of the plasmon absorption band was also observed, which was thought to attribute to the formation of Au nanorods. To account for this behavior, we believe the particle shape, and gradation in the depolarization factor of metals and in the volume fraction of the metallic (or dielectric) component should be expected to play an important role.

In a word, the sharp plasmon peak comes naturally from the existence of metal-dielectric interfaces. In the case of graded metallic films, there should be a broad band only, but no sharp peak. So, for the graded metal-dielectric composite film under present consideration, both the plasmon peak and the broad plasmon band should appear as predicted above.

In this section, we have studied the effective nonlinear optical response of a graded metal-dielectric composite film of anisotropic particles. Based on the MGA1 and MGA2 [Eqs. (5.3) and (5.4)], we derived the local electric field inside the film, and hence obtained the effective linear dielectric constant [Eq. (5.16)] and third-order nonlinear susceptibility [Eq. (5.18)] of the graded composite film.

In comparison with textbook formulae, our formulae [Eqs. (5.3) and (5.4)] only differ from the z-dependent volume fraction $p(z)$, in the sense that we could discuss the gradation which is perpendicular to the film and then leads to nonlinearity enhancement. As a matter of fact, the present results do not depend crucially on the particular form of the dielectric gradation profile $p(z)$ or the depolarization-factor gradation profile $L_z(z)$. The only requirement is that we must have a compositional or shape-dependent gradation to yield a broad plasmon band for the composite film. It should also be remarked that the opti-

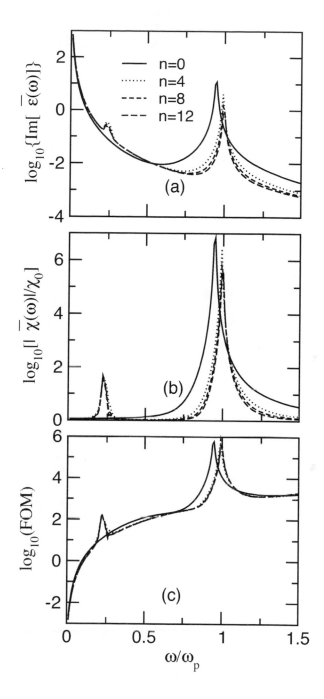

Figure 5.8. *Results for the MGA1 [Eq. (5.3)].* (a) Linear optical absorption $\mathrm{Im}[\bar{\varepsilon}(\omega)]$, (b) enhancement of the third-order optical nonlinearity $|\bar{\chi}(\omega)|/\chi_0$, and (c) FOM $\equiv |\bar{\chi}(\omega)|/\{\chi_0\mathrm{Im}[\bar{\varepsilon}(\omega)]\}$ versus the normalized incident angular frequency ω/ω_p for the gradation profile of the depolarization factor of dielectric particles $L_z^{(2)}(z) = (1/3)z^n$, for different n. Parameters: $p(z) = 0.85$, $\gamma/\omega_p = 0.01$, and $\varepsilon_2 = (3/2)^2$. After Ref. [117].

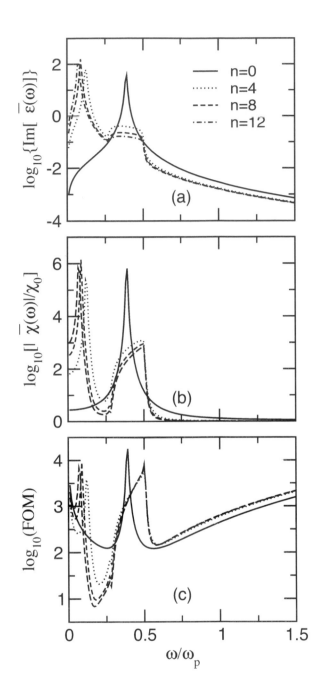

Figure 5.9. *Results for the MGA2 [Eq. (5.4)].*   (a) Linear optical absorption $\text{Im}[\bar{\varepsilon}(\omega)]$, (b) enhancement of the third-order optical nonlinearity $|\bar{\chi}(\omega)|/\chi_0$, and (c) FOM $\equiv |\bar{\chi}(\omega)|/\{\chi_0\text{Im}[\bar{\varepsilon}(\omega)]\}$ versus the normalized incident angular frequency ω/ω_p for the grada-tion profile of the depolarization factor of metallic particles $L_z^{(1)}(z) = (1/3)z^n$, for different n. Parameters: $p(z) = 0.85$, $\gamma/\omega_p = 0.01$, and $\varepsilon_2 = (3/2)^2$. After Ref. [117].

cal response of the graded structure depends on polarization of the incident light, because the incident optical field can always be resolved into two polarizations. However, a large nonlinearity enhancement occurs only when the electric field is parallel to the direction of the gradient [4], and the other polarization does not produce nonlinearity enhancement at all [4]. The nonlinear susceptibilities of both the parallel and perpendicular polarizations are related to the nonlinear phase shift which can be measured by using a z-scan method [4].

Following Roorda *et al.* [100], one may fabricate the film under the present discussion by using mega-electron-volt ion irradiation. Its third-order nonlinear susceptibility could also be measured by a degenerated four-wave-mixing method, which has been used for the Au/SiO$_2$ composite film [119]. It is of interest to extend the present theory to composites in which graded spherical particles are embedded in a host medium [13] to account for mutual interactions among graded particles.

To sum up, we have studied the effective linear dielectric constant and third-order non-linear susceptibility of a graded metal-dielectric composite film of anisotropic particles with weak nonlinearity by invoking the local field effects exactly within the Maxwell-Garnett theory. We have numerically demonstrated that this kind of film can serve as a novel optical material for producing a broad structure in both the linear and nonlinear response and an enhancement in the nonlinear response.

1.2. Second-Harmonic Generation

We consider a graded metal-dielectric composite film of thickness l_s with the gradation profile in the direction (z-direction) perpendicular to the film (Fig. 5.10). If one considers quadratic nonlinearities only, the local constitutive relation between the displacement field $\mathbf{D}(z)$ and the electric field $\mathbf{E}(z)$ in the static case is given by [76, 77]

$$D_i(z) = \sum_j \varepsilon_{ij}(z) E_j(z) + \sum_{jk} \chi_{ijk}(z) E_j(z) E_k(z), \ i = x, y, z, \qquad (5.20)$$

where $D_i(z)$ and $E_i(z)$ are the ith component of $\mathbf{D}(z)$ and $\mathbf{E}(z)$, respectively, and χ_{ijk} the SHG susceptibility. Here $\varepsilon_{ij}(z) = \varepsilon(z)\delta_{ij}$ denotes the linear dielectric constant, which is assumed for simplicity to be isotropic. Both $\varepsilon(z)$ and $\chi_{ijk}(z)$ are functions of z, as a result of the gradation profile in the z-direction.

If a monochromatic external field is applied, the nonlinearity in the system will generally generate local potentials and fields at all harmonic frequencies. For a finite frequency external electric field of the form

$$E_0 = E_0(\omega)e^{-i\omega t} + \text{c.c.}, \qquad (5.21)$$

the effective SHG susceptibility $\bar{\chi}_{2\omega}$ can be extracted by considering the volume average of the displacement field at the frequency 2ω in the inhomogeneous medium [16, 75, 76, 77].

Let $p(z)$ be the volume fraction of the metallic component in the graded film. To calculate $\varepsilon(z, \omega)$, we invoke the well-known Maxwell-Garnett approximation [115]

$$\frac{\varepsilon(z, \omega) - \varepsilon_2}{L_z \varepsilon(z, \omega) + (1 - L_z)\varepsilon_2} = p(z) \frac{\varepsilon_1(\omega) - \varepsilon_2}{L_z \varepsilon_1(\omega) + (1 - L_z)\varepsilon_2}, \qquad (5.22)$$

where $\varepsilon_1(\omega)$ and ε_2 are the linear dielectric constants of the metallic and dielectric components, respectively. Here, L_z is the depolarization factor describing the anisotropy of the metallic particles along the z-axis, with $0 < L_z < 1/3$ (or $1/3 < L_z < 1$) denoting prolate (or oblate) spheroids and $L_z = 1/3$ for spherical particles [91]. It is worth noting that prolate spheroidal particles can more easily be fabricated than oblate spheroidal particles in experiments using the method of ion irradiation (see, e.g., Ref. [100]). For completeness, we discuss both prolate and oblate spheroidal particles (Fig. 5.11). Implicit in Eq. (5.22) is the assumption that the major axes of the metallic particles are aligned perpendicular to the film. Experimentally, prolate spheroidal metallic particles can be made highly oriented along the direction of irradiated ions [100].

The dielectric function $\varepsilon_1(\omega)$ of the metallic component is taken to be the Drude form

$$\varepsilon_1(\omega) = 1 - \frac{\omega_p^2}{\omega(\omega + i\gamma)}, \tag{5.23}$$

where ω_p denotes the plasma frequency and γ the relaxation rate. For a z-dependent volume fraction profile $p(z)$, we can make use of the equivalent capacitance for capacitors in series to evaluate the effective perpendicular linear dielectric response $\bar{\varepsilon}(\omega)$ at a given frequency, i.e.,

$$\frac{1}{\bar{\varepsilon}(\omega)} = \frac{1}{L} \int_0^L \frac{dz}{\varepsilon(z,\omega)}. \tag{5.24}$$

The treatment of the effective linear response is analogous to the effective medium approximation for the thermal properties of graded films [92].

The calculation of the effective nonlinear optical response proceeds by applying the expressions derived in Ref.[75]. The effective SHG susceptibility $\bar{\chi}_{2\omega}(z)$ for the slice of the system at position z is given by [76, 77]

$$\bar{\chi}_{2\omega}(z) = p(z)\chi_{2\omega}\alpha(z,2\omega)[\alpha(z,\omega)]^2, \tag{5.25}$$

where $\alpha(z,\omega)$ denotes the local-field factor in a *linear* inhomogeneous system [75] which, for consistency with Eq. (5.22) in getting $\varepsilon(z,\omega)$, should also be determined by using the Maxwell-Garnett approach. The result is

$$\alpha(z,\omega) = \frac{\varepsilon_2}{\varepsilon_2[1 - L_z(1 - p(z))] + \varepsilon_1(\omega)L_z(1 - p(z))}. \tag{5.26}$$

In Eq. (5.25), $\chi_{2\omega}$ is the intrinsic SHG susceptibility of the metallic component. For simplicity, we assumed that the dielectric host is linear. The effective SHG susceptibility $\bar{\chi}_{2\omega}$ of the whole film can then be evaluated by a one-dimensional integral over the film thickness to give

$$\bar{\chi}_{2\omega} = \frac{1}{L} \int_0^L dz \bar{\chi}_{2\omega}(z) \left(\frac{E(z,2\omega)}{E_0}\right) \left(\frac{E(z,\omega)}{E_0}\right)^2, \tag{5.27}$$

where $E(z,\omega)$ denotes the volume average of the electric field within a layer at position z.

By virtue of the continuity of electric displacement, there is a relation

$$\varepsilon(z,\omega)E(z,\omega) = \bar{\varepsilon}(\omega)E_0. \tag{5.28}$$

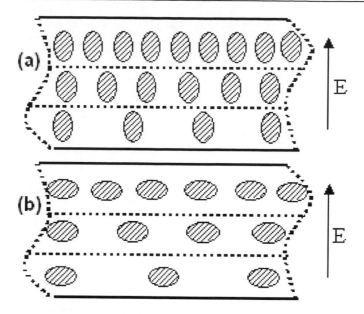

Figure 5.10. Schematic graph showing the geometry of a graded metal-dielectric composite film with a variation of volume fraction of (a) prolate and (b) oblate spheroidal metallic particles along z axis perpendicular to the film. The electric field E is parallel to the gradient along z axis. After Ref. [120].

Thus, we obtain

$$\bar{\chi}_{2\omega} = \frac{1}{L} \int_0^L dz \bar{\chi}_{2\omega}(z) \left(\frac{\bar{\varepsilon}(2\omega)}{\varepsilon(z, 2\omega)} \right) \left(\frac{\bar{\varepsilon}(\omega)}{\varepsilon(z, \omega)} \right)^2. \tag{5.29}$$

Equation (5.29), coupled with Eqs. (5.22) and (5.24), gives a compact expression for the effective SHG susceptibility in graded films.

For illustration, we take a model profile of metallic volume fraction of the form

$$p(z) = az^m, \tag{5.30}$$

where a and m are constants tuning the profile. Without loss of generality, the thickness L is set to unity, i.e., thickness is measured in units of L. Given values of m and a corresponding to a certain volume fraction p of metallic component in the film,

$$p = \frac{1}{L} \int_0^L p(z) dz. \tag{5.31}$$

For a given profile $p(z)$, if we *randomly* disperse the *same* amount of anisotropic metallic component in a film of thickness L, the effective SHG susceptibility χ_0 for the *random* film would be [76, 77]

$$\chi_0 = p\chi_{2\omega}\alpha(2\omega)[\alpha(\omega)]^2, \tag{5.32}$$

where the local-field factor $\alpha(\omega)$ is given by an expression similar to Eq. (5.26) as

$$\alpha(\omega) = \frac{\varepsilon_2}{\varepsilon_2[1 - L_z(1 - p)] + \varepsilon_1(\omega)L_z(1 - p)}. \tag{5.33}$$

By comparing the effective SHG susceptibility $\bar{\chi}_{2\omega}$ with χ_0 (see Fig. 5.12), we can see whether a graded film gives an enhanced SHG response, when compared to a non-graded film of random composite with the *same* volume fraction of metallic component.

We have carried out model numerical calculations to investigate the effects on the local field factors in Eq. (5.25) due to the gradation profile, the metallic particle shape, and the difference in the linear dielectric response between the two constituents. Figure 5.11 shows the effects of different degrees of anisotropy as specified by different values of the depolarization factor L_z. Figure 5.11 gives the real and imaginary parts of the effective linear dielectric constant $\bar{\varepsilon}(\omega)$ [Fig. 5.11(a) and (b)], and the real and imaginary parts of the effective SHG susceptibility $\bar{\chi}_{2\omega}$ [Fig. 5.11(c) and (d)] as a function of frequency ω/ω_p. Also shown is the modulus of $\bar{\chi}_{2\omega}/\chi_{2\omega}$ (see Fig. 5.11(e)). For Fig. 5.11 (and Fig. 5.12), we take the parameters $a = 0.8$ and $m = 1$, which correspond to the total volume fraction $p = 0.4$ according to Eq. (5.31). As L_z decreases, i.e., as the shape of the particles changes from oblate spheroid, to sphere, and then to prolate spheroid, the plasmon band becomes broader, and also shifts to lower frequencies, see Fig. 5.11(b). In Fig. 5.11(c)-(e), $\bar{\chi}_{2\omega}$ is normalized by the intrinsic SHG susceptibility of the metallic component $\chi_{2\omega}$, which is assumed to be a real and positive frequency-independent constant. In Fig. 5.11(c)-(e), there exists a frequency range in which a significantly enhanced SHG susceptibility results, when compared with $\chi_{2\omega}$. As L_z decreases, the frequency range becomes narrower and red-shifted to lower frequencies (see Fig. 5.11(c)-(e)).

It is also illustrative to compare the results in the presence of a gradation profile with that of a random composite film consisting of the same amount of metallic (nonlinear) component. In Fig. 5.12, we show the results for $\bar{\chi}_{2\omega}/\chi_0$, where χ_0 is given by Eq. (5.32). The results indicate that a gradation profile may not always enhance the SHG response. Generally speaking, one has to carefully make sure of the dielectric contrast, together with composition and gradation profiles, to achieve SHG enhancement in certain ranges of frequencies.

To further investigate the effects of a gradation profile, we consider a fixed volume fraction p of the nonlinear component. For a profile of the form $p(z) = az^m$, Eq. (5.31) implies that the parameters a and m are related by $p = a/(m+1)$. For a given value of p, we may adjust m (and a) to study the effects of different gradation profiles corresponding to the same value of p. Note that $m = 0$ refers to a non-graded random film. Figure 5.13 shows the results at the fixed frequency of $\omega/\omega_p = 0.2$ for different profiles characterized by the parameter m, for four different values of volume fraction $p = 0.1, 0.2, 0.3$, and 0.4. The linear responses can be enhanced to different extent with a gradation profile [see Fig. 5.13(a) and (b)]. The imaginary part of the linear dielectric response [Fig. 5.13(b)] shows a broad structure with frequency with a broad peak at frequencies at which the real part shows a sharp drop [Fig. 5.13(a)], expect for the system with the lowest concentration of nonlinear component. Figure 5.13(c)-(e) shows that the SHG responses are highly sensitive to the gradation profile. For the same concentration, one may tune the effective response by tuning the concentration profile. Note that for a range of m above $m = 0$ (see Fig. 5.13(e)), there is an increase in the SHG response with m for systems with $p > 0.1$, showing that a suitable gradation profile, which amounts to suitably placing a certain fraction of the the nonlinear component in the system, may provide an optimal SHG response for a given fraction of the nonlinear component. Our results show that for small total vol-

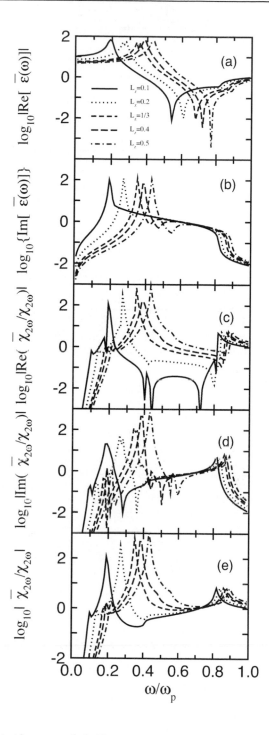

Figure 5.11. (a) $\mathrm{Re}[\bar{\varepsilon}(\omega)]$, (b) $\mathrm{Im}[\bar{\varepsilon}(\omega)]$ (linear optical absorption), (c) $\mathrm{Re}[\bar{\chi}_{2\omega}/\chi_{2\omega}]$, (d) $\mathrm{Im}[\bar{\chi}_{2\omega}/\chi_{2\omega}]$, and (e) Modulus of $\bar{\chi}_{2\omega}/\chi_{2\omega}$, versus the normalized incident angular frequency ω/ω_p for layer metal profile $p(z) = az^m$, for different L_z. Here $|\cdots|$ denotes the absolute value or modulus of $\cdots$. Parameters: $a = 0.8$, $m = 1$, $\gamma/\omega_p = 0.01$, and $\varepsilon_2 = (3/2)^2$. After Ref. [120].

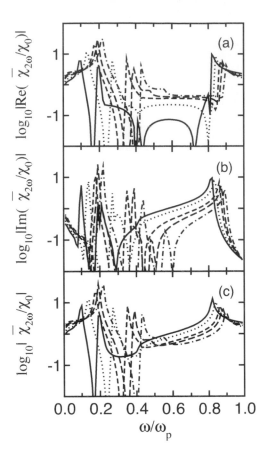

Figure 5.12. Same as Fig. 5.11(c)-(e), respectively, but (a) $\text{Re}[\bar{\chi}_{2\omega}/\chi_0]$, (b) $\text{Im}[\bar{\chi}_{2\omega}/\chi_0]$, and (c) Modulus of $\bar{\chi}_{2\omega}/\chi_0$. After Ref. [120].

ume fraction p ($p < 0.1$), a uniform profile or a profile that increases rapidly at small z is beneficial, while for moderate p ($p > 0.1$), there exists an optimal profile for SHG response. Note that for a given total volume fraction p, a gradation profile leads to non-trivial response in that the volume fraction $p(z)$ may be below the percolation threshold for same values of z and above the threshold for other values of z. Results of our model calculations show that a gradation profile is an additional means for tuning the local field effects.

Here we have investigated compositionally graded metal-dielectric composite films in which the fraction of the metallic component varies perpendicular to the film. Similarly, enhancement in the response was found for the polarization perpendicular to the film (i.e., parallel to the direction of the gradient), as a result of the continuity of the normal component of the displacement field. For the polarization parallel to the film, the physics is then governed by the continuity of the tangential component of the electric field [4]. Including nonlinear response into the consideration, it is expected that an enhanced SHG signal will result in a composite consisting of graded particles.

In summary, we have presented a formalism for evaluating the effective SHG susceptibility in compositionally graded metal-dielectric composite films with anistropically shaped

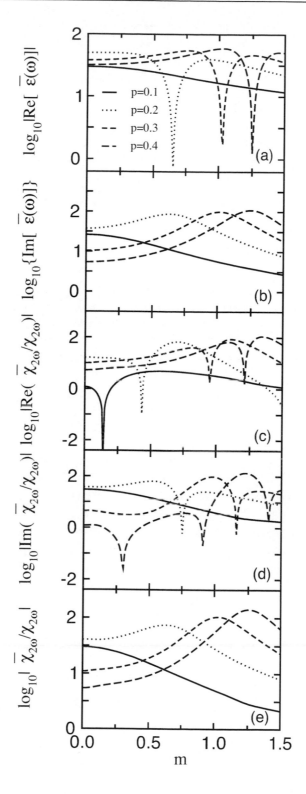

Figure 5.13. Same as Fig. 5.11, but versus m (dimensionless) for different total volume fraction p. Parameters: $L_z = 0.1$, $\omega/\omega_p = 0.2$, $\gamma/\omega_p = 0.01$, and $\varepsilon_2 = (3/2)^2$. After Ref. [120].

particles. We also carried out model numerical calculations to illustrate the effects of a gradation profile. It is found that the frequency-dependent response is highly sensitive to the degree of anisotropy of the particles. Suitably choosing the shape of the particles and the gradation profile for a given volume fraction of the particles, one may achieve tuning of an enhanced SHG susceptibility at specific frequency ranges.

2.   Graded Metal-Dielectric Films: Effects of Microstructure

The nonlinear optical properties of composite materials comprise an interesting field to study, since the properties may differ significantly from those of the constituents. One of the crucial elements for control of the linear and/or nonlinear optical responses is the microstructure of composites (e.g., see Refs. [44, 70] and references therein). For discussing the effect of microstructure, the usual two methods are the well-known Maxwell-Garnett approximation (MGA) [23, 24] and Bruggeman effective medium approximation (EMA) [25]. It is worth to noting that the MGA is an asymmetrical theory whereas the EMA is a symmetrical theory.

Let us consider a nonlinear graded film with width L, in which the dielectric particles of dielectric constant ε_2 is embedded in the host metal of $\varepsilon_1(\omega)$, and the volume fraction of the embedded particles $p_2(z)$ varies along $z-$axis. Here the gradient of gradation is in the direction perpendicular to the film, i.e., in $z-$axis. In this connection, the local constitutive relation between the displacement $\mathbf{D}$ and electric field $\mathbf{E}$ is given by

$$\mathbf{D}(z,\omega) = \varepsilon(z,\omega)\mathbf{E}(z,\omega) + \chi(z,\omega)|\mathbf{E}(z,\omega)|^2\mathbf{E}(z,\omega), \qquad (5.34)$$

where $\varepsilon(z,\omega)$ and $\chi(z,\omega)$ are respectively the linear dielectric constant and third-order nonlinear susceptibility of a layer inside the graded film. It is worth mentioning that both $\varepsilon(z,\omega)$ and $\chi(z,\omega)$ are gradation profiles as a function of position z.

Let us further assume that the weak nonlinearity condition is satisfied. That is, the contribution of the second term (nonlinear part $\chi(z,\omega)|\mathbf{E}(z,\omega)|^2$) in the right-hand side of Eq. (5.34) is much less than that of the first term (linear part $\varepsilon(z,\omega)$) [66]. Next, we restrict our discussion to the quasi-static approximation, i.e., $d/\lambda \leq 1$, where d is the characteristic size of the particle and λ is the wavelength of the incident light. In the quasi-static approximation, the whole graded structure can be regarded as an effective homogeneous one with effective (overall) linear dielectric constant $\bar{\varepsilon}(\omega)$ and effective (overall) third-order nonlinear susceptibility $\bar{\chi}(\omega)$. That is, $\bar{\varepsilon}(\omega)$ and $\bar{\chi}(\omega)$ is defined as [66]

$$\langle\mathbf{D}\rangle = \bar{\varepsilon}(\omega)\mathbf{E}_0 + \bar{\chi}(\omega)|\mathbf{E}_0|^2\mathbf{E}_0, \qquad (5.35)$$

where $\langle\cdots\rangle$ stands for the spatial average of $\cdots$, and $\mathbf{E}_0 = E_0\hat{e}_z$ the applied field along $z-$axis.

Now let us take into account the detailed microstructures in order to obtain $\bar{\varepsilon}(\omega)$ and $\bar{\chi}(\omega)$. First, we consider an asymmetrical microstructure [23, 24] in which one component constitutes inclusion particles while the other serves as a host. For such a microstructure, the MGA (Maxwell-Garnett approximation) is valid. In detail, for each layer of the graded metal-dielectric film, the effective dielectric constant $\varepsilon(z,\omega)$ is the solution of the MGA equation,

$$\frac{\varepsilon(z,\omega)-\varepsilon_1(\omega)}{\varepsilon(z,\omega)+2\varepsilon_1(\omega)} = p_2(z)\frac{\varepsilon_2-\varepsilon_1(\omega)}{\varepsilon_2+2\varepsilon_1(\omega)}, \tag{5.36}$$

On the other hand, we focus on a symmetrical microstructure in which the two sorts of particles can be exchanged, yielding, however, no effect on the effective dielectric constant $\varepsilon(z,\omega)$ of each layer. For treating such a symmetric microstructure, the EMA (Bruggeman effective medium approximation) [25] works for calculating $\varepsilon(z,\omega)$ by solving the self-consistent equation such that

$$(1-p_2(z))\frac{\varepsilon_1(\omega)-\varepsilon(z,\omega)}{\varepsilon_1(\omega)+2\varepsilon(z,\omega)} + p_2(z)\frac{\varepsilon_2-\varepsilon(z,\omega)}{\varepsilon_2+2\varepsilon(z,\omega)} = 0. \tag{5.37}$$

Regarding Eqs. (5.36) and (5.37), we should remark more. In fact, it is not possible to calculate $\varepsilon(z,\omega)$ exactly in terms of the layer dielectric profile $p_2(z)$. Nevertheless, to obtain an estimate of $\varepsilon(z,\omega)$, we can take a small volume element inside the layer, at a position z. Further, this small volume element can be seen as a composite where the dielectric particles are randomly embedded in the metallic component. Accordingly, the volume fraction of the dielectric particles is $p_2(z)$. In this regard, the above-mentioned MGA and EMA should be expected to hold well for computing $\varepsilon(z,\omega)$.

Owing to the simple graded structure, we can use the equivalent capacitance of series combination to calculate the linear response (i.e., optical absorption),

$$\frac{1}{\bar{\varepsilon}(\omega)} = \frac{1}{L}\int_0^L \frac{dz}{\varepsilon(z,\omega)}. \tag{5.38}$$

To investigate the nonlinear optical response, we first calculate local electric field $E(z,\omega)$ by means of the identity

$$\varepsilon(z,\omega)E(z,\omega) = \bar{\varepsilon}(\omega)E_0 \tag{5.39}$$

due to the virtue of the continuity of the electric displacement. In view of the existence of nonlinearity inside the graded film, the effective nonlinear response $\bar{\chi}(\omega)$ can be given by [66]

$$\bar{\chi}(\omega)E_0{}^4 = \langle\chi(z,\omega)|\mathbf{E}_{\text{lin}}(z)|^2\mathbf{E}_{\text{lin}}(z)^2\rangle, \tag{5.40}$$

where $\mathbf{E}_{\text{lin}}$ denotes the linear local electric field. Next, we take one step forward to express the effective nonlinear response as an integral over the film,

$$\bar{\chi}(\omega) = \frac{1}{L}\int_0^L dz\chi(z,\omega)\left|\frac{\bar{\varepsilon}(\omega)}{\varepsilon(z,\omega)}\right|^2\left(\frac{\bar{\varepsilon}(\omega)}{\varepsilon(z,\omega)}\right)^2. \tag{5.41}$$

For the following numerical calculations, we adopt a Drude-type dielectric function for metallic particles, namely,

$$\varepsilon_1(\omega) = 1 - \frac{\omega_p^2}{\omega(\omega+i\gamma)}, \tag{5.42}$$

where ω_p denotes the bulk plasmon frequency, and γ the damping constant. In addition, we set $\gamma = 0.01\omega_p$ (typical value for noble bulk metals) and $\varepsilon_2 = (3/2)^2$ (dielectric constant of glass).

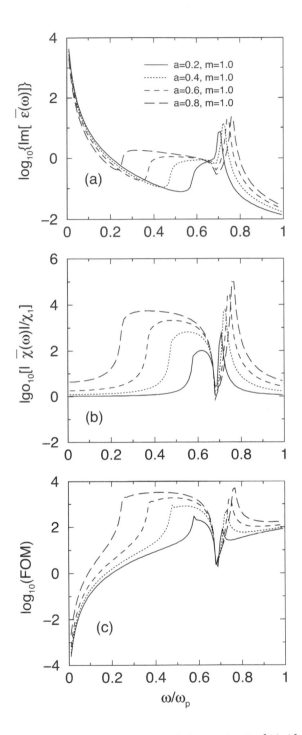

Figure 5.14. *Results of the MGA.* (a) Linear optical absorption $\mathrm{Im}[\bar{\varepsilon}(\omega)]$, (b) enhancement of the third-order optical nonlinearity $|\bar{\chi}(\omega)|/\chi_1$, and (c) FOM $\equiv |\bar{\chi}(\omega)|/\{\chi_1 \mathrm{Im}[\bar{\varepsilon}(\omega)]\}$ versus the normalized incident angular frequency ω/ω_p for layer dielectric profile $p_2(z) = az^m$. Parameters: $\gamma/\omega_p = 0.01$ and $\varepsilon_2 = (3/2)^2$.

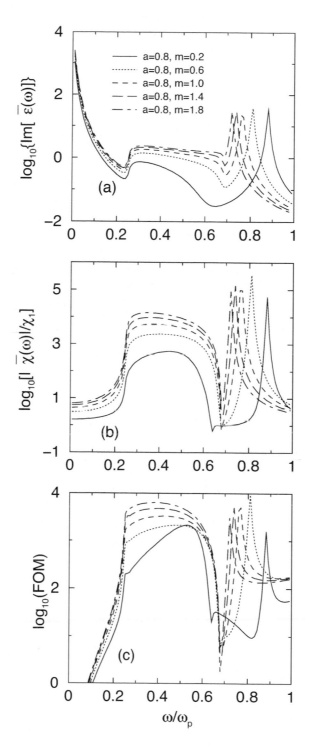

Figure 5.15. *Results of the MGA.* Same as Fig. 5.14, but for different m at $a = 0.8$.

Now we are in a position to do some numerical calculations in an attempt to discuss the effect of the above-mentioned microstructures. Set $\chi(z,\omega)$ to be a constant χ_1, so that we could emphasize the enhancement of the optical nonlinearity. Regarding the layer dielectric profile, take a power form $p_2(z) = az^m$. Without loss of generality, the layer width L is set to be unity.

In Figs. 5.14-5.17, we plot (a) the optical absorption $\sim \text{Im}[\bar{\varepsilon}(\omega)]$, (b) the modulus of the effective third-order optical nonlinearity enhancement $|\bar{\chi}(\omega)|/\chi_1$, and (c) the FOM $|\bar{\chi}(\omega)|/\{\chi_1\text{Im}[\bar{\varepsilon}(\omega)]\}$ as a function of the incident angular frequency ω, respectively. Here $\text{Im}[\cdots]$ means the imaginary part of $\cdots$.

It is well known that the MGA correctly predicts the surface plasmon resonance of bulk metal-dielectric composite, and this properties is shown as well in our model for a metal-dielectric film. Figures 5.14 and 5.15 show the optical properties based on the MGA. In Fig. 5.14, we display the effect of the coefficient a. When the layer dielectric profile $p_2(z)$ is taken into account, a broad resonant plasmon band is observed always. In other words, the broad band is caused to appear by the effect of the positional dependence of the dielectric or metal. Also, we find that increasing a causes the resonant band not only to be enhanced, but also red-shifted (namely, located at a lower frequency region). In a word, although the enhancement of the effective third-order optical nonlinearity is often accompanied with the appearance of the optical absorption, the FOM is still possible to be very attractive due to the presence of the positional dependence of the dielectric or metallic components. Moreover, it is worth noting that a prominent surface plasmon resonant peak appears at somewhat higher frequencies in addition to the surface plasmon band. As a increases, this peak is blue-shifted (i.e., locates at a higher frequency region) accordingly.

Similarly, Figure 5.15 displays the influence of m. It is apparent to see that the broad resonant plasmon band can be enhanced significantly by adjusting m. However, no distinct red-shift occurs for the plasmon band as m varies. In contrast, we notice that increasing m can make the surface plasmon resonant peak red-shifted.

For the symmetrical microstructure (EMA model), we also display the effects of a (Fig. 5.16) and m (Fig. 5.17), respectively. For this kind of microstructure, a plasmon band exists as well. However, the surface plasmon resonance becomes broad and weak, and the resonance peak disappears at a large volume fraction of metallic particles (e.g., $a = 0.2, m = 1.0$). This is different from the MGA prediction that whatever the volume fraction of metal component is there always exists a sharp resonance. That is, such a difference is caused by introducing the two different (asymmetrical and symmetrical) microstructures. Moreover, it is well known that there is a percolation threshold predicted by the EMA, at which the properties of the metal-dielectric composite change significantly. For metal-dielectric composites the percolation threshold is $p_2(z) = 2/3$ [i.e., $1 - p_2(z) = 1/3$] at which the conductivity of the composite becomes nonzero. In other words, as $1 - p_2(z) > 1/3$ the composite behaves as a metal rather that a dielectric. Thus, in Figs. 5.16 and 5.17, there is no apparent resonance peak especially when the volume fraction of the metal is large. Also, the nonlinearity enhancement is unimpressive and the FOM is generally small. Nevertheless, the plasmon band always exists, too.

For the Maxwell-Garnett model, the microstructure of interest should be asymmetrical. That is, the dielectric particles are surrounded by the metallic component. In other words, the dielectric particles are randomly dispersed in a metallic host so that the dielec-

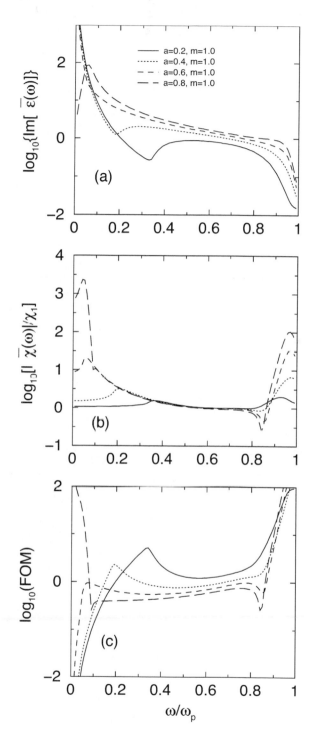

Figure 5.16. *Results of the EMA*. Others are the same as Fig. 5.14.

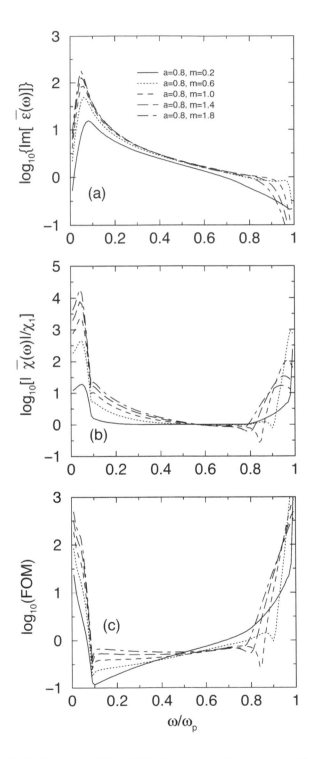

Figure 5.17. *Results of the EMA.* Others are the same as Fig. 5.15.

tric particles cannot touch each other. On the other hand, for the Bruggeman model, the microstructure should be symmetrical. That is, both the dielectric particles and metallic particles are mixed randomly. It is not possible to distinguish the embedded component with the host medium. It is known that the Maxwell-Garnett model (without gradation) can predict a sharp plasmon resonance peak whereas the Bruggeman model will give a broad plasmon band. For the Maxwell-Garnett model with gradation, a broad resonant plasmon band is observed, see Figs. 5.14-5.15. It is because the inhomogeneity due to gradation leads to a further broadening of the plasmon resonance peak. In more detail, as a or m increases, see Fig. 5.14 or Fig. 5.15, the resonant frequency takes on values within a broader range across the film, and hence leads to a broad plasmon band. The further broadening in the plasmon band also occurs in the Bruggeman model.

We have discussed the effective linear and nonlinear optical responses of a compositionally graded metal-dielectric film in an attempt to study the effect of microstructure. For two asymmetrical and symmetrical microstructures, we used the MGA and EMA to calculate the effective responses, respectively.

The appearance of plasmon bands is interesting, and comes about from the gradual changes in the volume fraction of the metallic component in one direction, i.e. like a one-dimensional tight binding band of surface plasmon modes at each layer. As a matter of fact, the present results do not depend crucially on the particular form of the layer dielectric profile $p(z)$. However, the microstructure can significantly affect the linear and nonlinear optical response. As the volume fraction of the metal component increases, the MGA (or EMA) predicts a sharp (or broad and weak) resonance. However, a plasmon band was observed always, regardless of the detailed microstructure. To obtain such results, the only requirement is that one must have a composition-dependent layer inside the graded film.

In fact, a direct numerical evaluation of the linear problem giving E_{lin} can give an independent check for the validity of Eqs. (5.38) and (5.39). In doing so, one needs to invoke a random resistor network for the genuine random composite.

To calculate $\varepsilon(z, \omega)$ for an asymmetrical microstructure, we used the MGA [Eq. (5.36)] for which the metal (or dielectric) component serves as a host (or inclusion). Inversely, we could see the metal (or dielectric) component as an inclusion (or host), and use the same form as Eq. (5.36) by exchanging $\varepsilon_1(\omega)$ and ε_2, and $p_2(z)$ and $1 - p_2(z)$. In doing so, as $\varepsilon_1(\omega) > \varepsilon_2$ the former MGA [Eq. (5.36)] always gives an upper bound while the latter offers a lower bound, and vice versa. But the exact result must lie between the two bounds, as already pointed out by Hashin and Shtrikman [116].

To sum up, we have studied a compositionally graded metal-dielectric film by investigating two asymmetrical and symmetrical microstructures, respectively, and found that the effective linear and nonlinear optical responses are sensitive to the microstructure. Thus, it is possible to gain a large nonlinearity enhancement and optimal FOM by choosing an appropriate microstructure in compositionally graded metal-dielectric films.

3.   Composites of Graded Particles

The composite media consisting of graded inclusions can be more useful and interesting than those of homogeneous inclusions. Although various theories have been established to investigate the optical and dielectric properties of the composite media of homogeneous inclusions [62], they fail to deal with the inhomogeneous composites of graded inclusions directly. Recently, a first-principles approach [63, 121] and a differential effective dipole approximation [64, 65] have been presented in order to investigate the dielectric response of graded materials.

The problem becomes more complicated by the presence of nonlinearity in realistic composites. Besides inhomogeneity, such nonlinearity plays also an important role in the effective material properties of composite media [10, 55, 66, 67, 70, 73]. It is thus necessary to establish a new theory to study the effective nonlinear properties of graded composite media. In fact, the introduction of dielectric gradation profiles in nonlinear composites is able to provide an alternative way to control the local field fluctuation, and hence let us obtain the desired effective nonlinear response.

In fact, the previous one-shell model [74] and multi-shell model [69], which were used to study the effective nonlinear optical property, can be seen as an initial model of graded inclusions. Beblow, we will put forth a nonlinear differential effective dipole approximation (NDEDA) to investigate the effective linear and nonlinear dielectric properties of composite media containing a very small volume fraction of nonlinear graded spherical particles (inclusions). For such particles, the linear and nonlinear physical properties will continuously vary along their radius. For more details on graded particles, please refer to Appendices C-F.

3.1. Third-Order Nonlinearity

A. Model and definition of effective linear and nonlinear responses

Let us consider a nonlinear composite system, in which identical graded spherical inclusions with radius a, are randomly embedded in a linear host medium of dielectric constant ε_2. The local constitutive relation between the displacement ($\mathbf{D}$) and the electric field ($\mathbf{E}$) inside the graded particle is given by

$$\mathbf{D} = \varepsilon(r)\mathbf{E} + \chi(r)|\mathbf{E}|^2\mathbf{E}, \tag{5.43}$$

where $\varepsilon(r)$ and $\chi(r)$ are, respectively, the linear dielectric constant and third-order nonlinear susceptibility. Note both $\varepsilon(r)$ and $\chi(r)$ are radial functions. Here we assume that the weak nonlinearity condition is satisfied [66]. In other words, the contribution of the second (nonlinear) part $[\chi_s(r)|\mathbf{E}|^2]$ in the right-hand side of Eq. (5.43) is much less than that of the first (linear) part $\varepsilon(r)$. We restrict further our discussion to the quasi-static approximation, under which the whole composite medium can be regarded as an effective homogeneous one with effective linear dielectric constant ε_e and effective third-order nonlinear susceptibility χ_e. To show the definitions of ε_e and χ_e, we have [66]

$$\langle\mathbf{D}\rangle = \varepsilon_e\mathbf{E}_0 + \chi_e|\mathbf{E}_0|^2\mathbf{E}_0, \tag{5.44}$$

where $\langle\cdots\rangle$ represents the spatial average, and $\mathbf{E}_0 = E_0\mathbf{e}_z$ is the external applied field along z axis.

The effective linear dielectric constant ε_e is given by

$$\varepsilon_e\mathbf{E}_0 = \frac{1}{V}\int_V \varepsilon_i\mathbf{E}_{\text{lin},i}dV = f\langle\varepsilon(r)\mathbf{E}_{\text{lin},1}\rangle + (1-f)\varepsilon_2\langle\mathbf{E}_{\text{lin},2}\rangle, \tag{5.45}$$

where f is the volume fraction of the graded particles and the subscript stands for the linear local field [i.e., obtained for the same system but with $\chi(r) = 0$].

In view of the existence of nonlinearity inside the graded particles, χ_e can then be written as [66, 114]

$$\chi_e E_0^4 = \frac{1}{V} \int_V \chi_i |\mathbf{E}|_{\text{lin},i}^2 \mathbf{E}_{\text{lin},i}^2 dV = \frac{1}{V} \int_{\Omega_i} \chi(r) |\mathbf{E}|_{\text{lin},1}^2 \mathbf{E}_{\text{lin},1}^2 dV = f \langle \chi(r) |\mathbf{E}|_{\text{lin},1}^2 \mathbf{E}_{\text{lin},1}^2 \rangle. \quad (5.46)$$

In the next section, we will develop a NDEDA (nonlinear differential effective dipole approximation), in an attempt to derive the equivalent linear dielectric constant $\bar{\varepsilon}(a)$ and third-order nonlinear susceptibility $\bar{\chi}(a)$ of the nonlinear graded inclusions. Then, the effective linear dielectric constant and third-order nonlinear susceptibility of the composite media of nonlinear graded inclusions will be derived accordingly in the dilute limit.

B. Nonlinear differential effective dipole approximation

To establish the NDEDA, we first mimic the gradation profile by a multi-shell construction. That is, we build up the dielectric profile by adding shells gradually [64]. We start with an infinitesimal spherical core with linear dielectric constant $\varepsilon(0)$ and third-order nonlinear susceptibility $\chi(0)$, and keep on adding spherical shells with linear dielectric constant $\varepsilon(r)$ and third-order nonlinear susceptibility $\chi(r)$ at radius r, until $r = a$ is reached. At radius r, the inhomogeneous spherical particle with space-dependent dielectric gradation profiles $\varepsilon(r)$ and $\chi(r)$ can be replaced by a *homogenous* sphere with the equivalent dielectric properties $\bar{\varepsilon}(r)$ and $\bar{\chi}(r)$. Here the *homogeneous* sphere should induce the same dipole moment as the original inhomogeneous sphere.

Next, we add to the sphere a spherical shell of infinitesimal thickness dr, with dielectric constant $\varepsilon(r)$ and nonlinear susceptibility $\chi(r)$. In this sense, the coated inclusions is composed of a spherical core with radius r, linear dielectric constant $\bar{\varepsilon}(r)$ and nonlinear susceptibility $\bar{\chi}(r)$, and a shell with outermost radius $r + dr$, linear dielectric constant $\varepsilon(r)$ and nonlinear susceptibility $\chi(r)$. Since these coated inclusions with a very small volume fraction are randomly embedded in a linear host medium, under the quasi-static approximation, we can readily obtain the linear electric potentials in the core, shell and host medium by solving the Laplace equation [69]

$$
\begin{aligned}
\phi_c &= -E_0 A R \cos\theta, & R < r, \\
\phi_s &= -E_0 \left(BR - \frac{Cr^3}{R^2} \right) \cos\theta, & r < R < r + dr, \\
\phi_h &= -E_0 \left(R - \frac{D(r+dr)^3}{R^2} \right) \cos\theta, & R > r + dr,
\end{aligned}
\quad (5.47)
$$

where

$$
\begin{aligned}
A &= \frac{9\varepsilon_2 \varepsilon(r)}{Q}, \qquad B = \frac{3\varepsilon_2 [\bar{\varepsilon}(r) + 2\varepsilon(r)]}{Q}, \qquad C = \frac{3\varepsilon_2 [\bar{\varepsilon}(r) - \varepsilon(r)]}{Q}, \\
D &= \frac{[\varepsilon(r) - \varepsilon_2][\bar{\varepsilon}(r) + 2\varepsilon(r)] + \lambda[\varepsilon_2 + 2\varepsilon(r)][\bar{\varepsilon}(r) - \varepsilon(r)]}{Q},
\end{aligned}
$$

with interfacial parameter $\lambda \equiv [r/(r+dr)]^3$, and

$$Q = [\varepsilon(r) + 2\varepsilon_2][\bar{\varepsilon}(r) + 2\varepsilon(r)] + 2\lambda[\varepsilon(r) - \varepsilon_2][\bar{\varepsilon}(r) - \varepsilon(r)].$$

The effective (overall) linear dielectric constant of the system is determined by the dilute-limit expression [71]

$$\varepsilon_e = \varepsilon_2 + 3p\varepsilon_2 D, \tag{5.48}$$

where p is the volume fraction of graded particles with radius r. The equivalent dielectric constant $\bar{\varepsilon}(r+dr)$ for the graded particles with radius $r+dr$ can be obtained self-consistently by the vanishing of the dipole factor D by replacing ε_2 with $\bar{\varepsilon}(r+dr)$. Taking the limit $dr \to 0$ and keeping to the first order in dr, we obtain

$$
\begin{aligned}
\bar{\varepsilon}(r+dr) &= \varepsilon(r) + 3\bar{\varepsilon}(r)\lambda \cdot \frac{\bar{\varepsilon}(r) - \varepsilon(r)}{\bar{\varepsilon}(r)(1-\lambda) + \varepsilon(r)(2+\lambda)} \\
&= \bar{\varepsilon}(r) - \frac{\bar{\varepsilon}(r) - \varepsilon(r)}{r} \cdot \left[3 + \frac{\bar{\varepsilon}(r) - \varepsilon(r)}{\varepsilon(r)} \right] dr.
\end{aligned}
\tag{5.49}
$$

Thus, we have the differential equation for the equivalent dielectric constant $\bar{\varepsilon}(r)$ as [64]

$$\frac{d\bar{\varepsilon}(r)}{dr} = \frac{[\varepsilon(r) - \bar{\varepsilon}(r)] \cdot [\bar{\varepsilon}(r) + 2\varepsilon(r)]}{r\varepsilon(r)}. \tag{5.50}$$

Note that Eq. (5.112) is just the Tartar formula, derived for assemblages of spheres with varying radial and tangential conductivity [43].

Next, we speculate on how to derive the equivalent nonlinear susceptibility $\bar{\chi}(r)$. After applying Eq. (5.106) to the coated particles with radius $r+dr$, we have

$$\bar{\chi}(r+dr)\frac{\langle|\mathbf{E}|^2\mathbf{E}^2\rangle_{R\leq r+dr}}{|\mathbf{E}_0|^2\mathbf{E}_0^2} = \lambda\bar{\chi}(r)\frac{\langle|\mathbf{E}|^2\mathbf{E}^2\rangle_{R\leq r}}{|\mathbf{E}_0|^2\mathbf{E}_0^2} + (1-\lambda)\frac{\langle\chi(r)|\mathbf{E}|^2\mathbf{E}^2\rangle_{r<R\leq r+dr}}{|\mathbf{E}_0|^2\mathbf{E}_0^2}. \tag{5.51}$$

As $dr \to 0$, the left-hand side of the above equation admits

$$
\begin{aligned}
\bar{\chi}(r+dr)\frac{\langle|\mathbf{E}|^2\mathbf{E}^2\rangle_{R\leq r+dr}}{|\mathbf{E}_0|^2\mathbf{E}_0^2} &= \bar{\chi}(r+dr)\left|\frac{3\varepsilon_2}{\bar{\varepsilon}(r+dr)+2\varepsilon_2}\right|^2 \left(\frac{3\varepsilon_2}{\bar{\varepsilon}(r+dr)+2\varepsilon_2}\right)^2 \\
&= \bar{\chi}(r)|K|^2 K^2 - dr\bar{\chi}(r)|K|^2 K^2 \\
&\quad \times \left[\frac{3d\bar{\varepsilon}(r)/dr}{2\varepsilon_2+\bar{\varepsilon}(r)} + \left(\frac{d\bar{\varepsilon}(r)/dr}{2\varepsilon_2+\bar{\varepsilon}(r)}\right)^* \right] \\
&\quad + |K|^2 K^2 \frac{d\bar{\chi}(r)}{dr} \cdot dr,
\end{aligned}
\tag{5.52}
$$

with $K = (3\varepsilon_2)/[\bar{\varepsilon}(r)+2\varepsilon_2]$. The first part of the right-hand side of Eq. (5.113) is written as

$$\lambda\frac{\bar{\chi}(r)\langle|\mathbf{E}|^2\mathbf{E}^2\rangle_{R\leq r}}{|\mathbf{E}_0|^2\mathbf{E}_0^2} = \bar{\chi}(r)|K|^2 K^2 \left[1 + (6y + 2y^* - 3)\frac{dr}{r} \right], \tag{5.53}$$

where

$$y = \frac{[\varepsilon(r) - \varepsilon_2][\bar{\varepsilon}(r) - \varepsilon(r)]}{\varepsilon(r)[\bar{\varepsilon}(r) + 2\varepsilon_2]}.$$

The second part of the right-hand side of Eq. (5.113) has the form [71]

$$
\begin{aligned}
(1-\lambda)\frac{\langle\chi(r)|\mathbf{E}|^2\mathbf{E}^2\rangle_{r<R\leq r+dr}}{|\mathbf{E}_0|^2\mathbf{E}_0^2} &= \frac{3\chi(r)}{5r}dr|z|^2 z^2 (5 + 18x^2 + 18|x|^2 + 4x^3 \\
&\quad + 12x|x|^2 + 24|x|^2 x^2),
\end{aligned}
\tag{5.54}
$$

where

$$x = \frac{\bar{\varepsilon}(r) - \varepsilon(r)}{\bar{\varepsilon}(r) + 2\varepsilon(r)} \quad \text{and} \quad z = \frac{\varepsilon_2[\bar{\varepsilon}(r) + 2\varepsilon(r)]}{\varepsilon(r)[\bar{\varepsilon}(r) + 2\varepsilon_2]}.$$

Substituting Eqs. (5.114), (5.115) and (5.116) into Eq. (5.113), we have a differential equation for the equivalent nonlinear susceptibility $\bar{\chi}(r)$, namely,

$$\frac{d\bar{\chi}(r)}{dr} = \bar{\chi}(r) \left[\frac{3d\bar{\varepsilon}(r)/dr}{2\varepsilon_2 + \bar{\varepsilon}(r)} + \left(\frac{d\bar{\varepsilon}(r)/dr}{2\varepsilon_2 + \bar{\varepsilon}(r)} \right)^* \right] + \bar{\chi}(r) \frac{6y + 2y^* - 3}{r} + \frac{3\chi(r)}{5r}$$

$$\times \left| \frac{\bar{\varepsilon}(r) + 2\varepsilon(r)}{3\varepsilon(r)} \right|^2 \left(\frac{\bar{\varepsilon}(r) + 2\varepsilon(r)}{3\varepsilon(r)} \right)^2 (5 + 18x^2 + 18|x|^2 + 4x^3$$

$$+ 12x|x|^2 + 24|x|^2x^2). \tag{5.55}$$

So far, the equivalent $\bar{\varepsilon}(r)$ and $\bar{\chi}(r)$ of graded spherical particles of radius r can be calculated, at least numerically, by solving the differential equations Eqs. (5.112) and (5.117), as long as $\varepsilon(r)$ (dielectric-constant gradation profile) and $\chi(r)$ (nonlinear-susceptibility gradation profile) are given. Here we would like to mention that, even though $\chi(r)$ is independent of r, the equivalent $\bar{\chi}(r)$ should still be dependent on r because of $\varepsilon(r)$ as a function of r. Moreover, for both $\varepsilon(r) = \varepsilon_1$ and $\chi(r) = \chi_1$ (i.e., they are both constant and independent of r), Eqs. (5.112) and (5.117) will naturally reduce to the solutions $\bar{\varepsilon}(r) = \varepsilon_1$ and $\bar{\chi}(r) = \chi_1$.

To obtain $\bar{\varepsilon}(r = a)$ and $\bar{\chi}(r = a)$, we integrate Eqs. (5.112) and (5.117) numerically at given initial conditions $\bar{\varepsilon}(r \to 0)$ and $\bar{\chi}(r \to 0)$. Once $\bar{\varepsilon}(r = a)$ and $\bar{\chi}(r = a)$ are calculated, we can take one step forward to work out the effective linear and nonlinear responses ε_e and χ_e of the whole composite in the dilute limit, i.e. [66],

$$\varepsilon_e = \varepsilon_2 + 3\varepsilon_2 f \frac{\bar{\varepsilon}(r = a) - \varepsilon_2}{\bar{\varepsilon}(r = a) + 2\varepsilon_2}, \tag{5.56}$$

and

$$\chi_e = f\bar{\chi}(r = a) \left| \frac{3\varepsilon_2}{\bar{\varepsilon}(r = a) + 2\varepsilon_2} \right|^2 \left(\frac{3\varepsilon_2}{\bar{\varepsilon}(r = a) + 2\varepsilon_2} \right)^2. \tag{5.57}$$

C. Exact solution for power-law gradation profiles

Based on the first-principles approach, we have found that, for a power-law dielectric gradation profile, i.e., $\varepsilon(r) = A(r/a)^n$, the potential in the graded inclusions and the host medium can be exactly given by [121]

$$\phi_i(r) = -\xi_1 E_0 r^s \cos\theta, \quad r < a,$$

$$\phi_h(r) = -E_0 r \cos\theta + \frac{\xi_2}{r^2} E_0 \cos\theta, \quad r > a, \tag{5.58}$$

where the coefficients ξ_1 and ξ_2 have the form

$$\xi_1 = \frac{3a^{1-s}\varepsilon_2}{sA + 2\varepsilon_2} \quad \text{and} \quad \xi_2 = \frac{sA - \varepsilon_2}{sA + 2\varepsilon_2} a^3,$$

and s is given by

$$s = \frac{1}{2} \left[\sqrt{9 + 2n + n^2} - (1 + n) \right].$$

The local electric field inside the graded inclusions can be derived from the potential $E = -\nabla\phi$,

$$
\begin{aligned}
\mathbf{E}_i &= \xi_1 E_0 r^{s-1}(s\cos\theta\mathbf{e}_r - \sin\theta\mathbf{e}_\theta) = \xi_1 E_0 r^{s-1}\{(s-1)\cos\theta\sin\theta\cos\phi\mathbf{e}_x \\
&\quad + (s-1)\cos\theta\sin\theta\sin\phi\mathbf{e}_y + [(s-1)\cos^2\theta + 1]\mathbf{e}_z\},
\end{aligned} \tag{5.59}
$$

where $\mathbf{e}_r$, $\mathbf{e}_\theta$, and $\mathbf{e}_x$, $\mathbf{e}_y$ and $\mathbf{e}_z$ are unix vectors in spherical coordinates and in Cartesian coordinates. In the dilute limit, from Eq. (5.45), we can obtain the effective linear dielectric constant as follows

$$
\begin{aligned}
\varepsilon_e &= \varepsilon_2 + \frac{1}{VE_0}\int_{\Omega_i}[A(r/a)^n - \varepsilon_2]\mathbf{e}_z\cdot\mathbf{E}_i dV \\
&= \varepsilon_2 + 3\varepsilon_2 f\frac{2+s}{sA+2\varepsilon_2}\left(\frac{A}{2+n+s} - \frac{\varepsilon_2}{2+s}\right).
\end{aligned} \tag{5.60}
$$

On the other hand, the substitution of Eq. (5.59) into Eq. (5.106) yields

$$
\begin{aligned}
\chi_e &= \frac{1}{V}\int_{\Omega_i}\chi(r)|\xi_1|^2\xi_1^2(s^2\cos^2\theta + \sin^2\theta)^2 r^{4s-2}\sin\theta dr d\theta d\phi \\
&= \frac{f}{5a^3}|\xi_1|^2\xi_1^2(8+4s+3s^4)\cdot\int_0^a\chi(r)r^{4s-2}dr.
\end{aligned} \tag{5.61}
$$

For example, for a linear profile of $\chi(r)$, i.e., $\chi(r) = k_1 + k_2\cdot r/a$, Eq. (5.61) leads to

$$
\chi_e = \frac{f}{20}\left|\frac{3\varepsilon_2}{sA+2\varepsilon_2}\right|^2\left(\frac{3\varepsilon_2}{sA+2\varepsilon_2}\right)^2(8+4s^2+3s^4)\left(\frac{k_2}{s} + \frac{4k_1}{4s-1}\right). \tag{5.62}
$$

In addition, for a power-law profile of $\chi(r)$, namely, $\chi(r) = k_1(r/a)^{k_2}$, Eq. (5.61) produces

$$
\chi_e = \frac{f}{5}\left|\frac{3\varepsilon_2}{sA+2\varepsilon_2}\right|^2\left(\frac{3\varepsilon_2}{sA+2\varepsilon_2}\right)^2 k_1\left(\frac{8+4s^2+3s^4}{k_2-1+4s}\right). \tag{5.63}
$$

We are now in a position to evaluate the NDEDA. For the comparison between the first-principles approach and the NDEDA, we first perform numerical calculations for the case where the dielectric constant exhibits power-law gradation profiles $\varepsilon(r) = A(r/a)^n$, while the third-order nonlinear susceptibility shows two model gradation profiles: (a) linear profile $\chi(r) = k_1 + k_2\cdot r/a$, and (b) power-law profile $\chi(r) = k_1(r/a)^{k_2}$. Without loss of generality, we take $\varepsilon_2 = 1$ and $a = 1$ for numerical calculations. The fourth-order Runge-Kutta algorithm is adopted to integrate the differential equations [Eqs. (5.112) and (5.117)] with step size 0.01. Meanwhile, the initial core radius is set to be 0.001. It was verified that this step size guarantees accurate numerics.

In Fig. 5.18, the effective linear dielectric constant (ε_e) is plotted as a function of A for various indices n. It is shown that ε_e exhibits a monotonic increase for increasing A (and decreasing n). This can be understood by using the equivalent dielectric constant $\bar{\varepsilon}(r = a)$ which increases as A increases (n decreases). Moreover, the excellent agreement between the NDEDA [Eq. (5.112)] and the first-principles approach [Eq. (5.60)] is shown as well.

Next, the effective third-order nonlinear susceptibility (χ_e) is plotted as a function of A for the linear gradation profile $\chi(r) = k_1 + k_2\cdot r/a$ (Fig. 5.19), and for the power-law profile

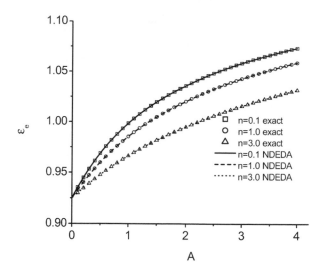

Figure 5.18. The effective linear dielectric constant ε_e versus A for the power-law dielectric gradation profile $\varepsilon(r) = A(r/a)^n$ in the dilute limit $f = 0.05$. Lines: numerical results from the NDEDA [Eq. (5.112)]; Symbols: exact results [Eq. (5.60)]. After Ref. [122].

$\chi(r) = k_1(r/a)^{k_2}$ (Fig. 5.20). We find that the effective nonlinear susceptibility decreases for increasing A. The reason is that, as mentioned above, for larger A, the graded inclusions possess larger equivalent dielectric constant, and the local field inside the nonlinear inclusions will become more weak, which results in a weaker effective nonlinear susceptibility (χ_e). In addition, increasing n leads generally to increasing χ_e, and such a trend is clearly observed at large A. Again, we obtain the excellent agreement between the first-principles approach [Eqs. (5.62) and (5.63)] and the NDEDA [Eqs. (5.112) and (5.117)].

In what follows, we investigate the surface plasmon resonance effect on the metal-dielectric composite. We adopt the Drude-like dielectric constant for graded metal particles, namely,

$$\varepsilon(r) = 1 - \frac{\omega_p^2(r)}{\omega[\omega + i\gamma(r)]}, \tag{5.64}$$

where $\omega_p(r)$ and $\gamma(r)$ are the radius-dependent plasma frequency and damping coefficient, respectively. For the sake of simplicity, set $\chi(r) = \chi_1$ to be independent of r, in an attempt to emphasize the enhancement of the effective optical nonlinearity, and $\varepsilon_2 = 1.77$ (the dielectric constant of water). We assume further $\omega_p(r)$ to be

$$\omega_p(r) = \omega_p\left(1 - k_\omega \cdot \frac{r}{a}\right), \quad r < a. \tag{5.65}$$

This form is quite physical for $k_\omega > 0$, since the center of grains can be better metallic so that $\omega_p(r)$ is larger, while the boundary of the grain may be poorer metallic so that $\omega_p(r)$ is much smaller. Such the variation can also appear because of the temperature effect [123]. For small particles, we have the radius-dependent $\gamma(r)$ as [84]

$$\gamma(r) = \gamma(\infty) + \frac{k_\gamma}{r/a}, \quad r < a, \tag{5.66}$$

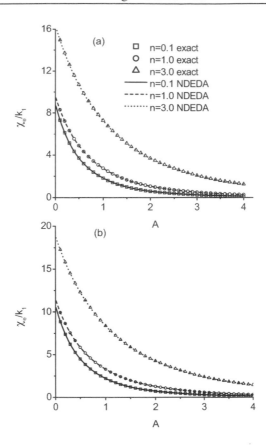

Figure 5.19. The effective third-order nonlinear susceptibility χ_e versus A for power-law dielectric-constant gradation profile $\varepsilon(r) = A(r/a)^n$ and linear nonlinear-susceptibility gradation profile $\chi(r) = k_1 + k_2 \cdot r/a$ with (a) $k_1 = 1$ and $k_2 = 1$, and (b) $k_1 = 2$ and $k_2 = 3$. Lines: numerical results from the NDEDA [Eqs. (5.112) and (5.117)]; Symbols: exact results [Eq. (5.61)]. After Ref. [122].

where $\gamma(\infty)$ stands for the damping coefficient in the bulk material. Here k_γ is a constant which is related to the Fermi velocity v_F. In this case, the exact solution being predicted by a first-principles approach is absent. Fortunately, we can resort to the NDEDA instead.

In Fig. 5.21, we plot the optical absorption [$\sim \text{Im}(\varepsilon_e)$], the modulus of the effective third-order optical nonlinearity enhancement ($|\chi_e|/\chi_1$) and the figure of merit ($|\chi_e|/\text{Im}(\varepsilon_e)$) versus the incident angular frequency ω. For the case of the homogeneous particles, i.e., $k_\omega = 0$, there is a single sharp peak at $\omega \approx 0.5\omega_p$, corresponding to the surface plasmon resonance, as expected. However, for the case of the graded particles, i.e. $k_\omega \neq 0$, besides a sharp peak, a broad continuous resonant band in the high-frequency region is apparently observed. The position of the sharp peak can be estimated from the resonant condition $\text{Re}[\bar{\varepsilon}(r = a)] + 2\varepsilon_2 = 0$, while the broad continuous spectrum is indeed a salient result of the gradation profile. More exactly, the broad spectrum results from the effect of the radius-dependent plasma frequency. In Ref. [74], we found that, when the shell model is taken into account, a broad continuous spectrum should be expected to occur around the large

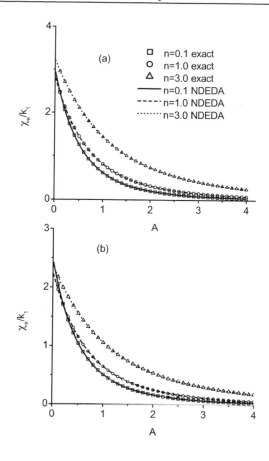

Figure 5.20. Same as Fig. 5.19, but for power-law nonlinear-susceptibility gradation profile $\chi(r) = k_1(r/a)^{k_2}$. After Ref. [122].

pole in the spectral density function. In fact, the graded particles under consideration can be regarded as a certain limit of multi shells, which thus should yield the broader spectra in $\text{Im}(\varepsilon_e)$, $|\chi_e|/\chi_1$ as well as $|\chi_e|/\text{Im}(\varepsilon_e)$. In addition, we note that increasing k_ω makes both the surface plasmon frequency and the center of the resonant bands red-shifted. In particular, the resonant bands can become more broad due to strong inhomogeneity of the particles. From the figure, we conclude that, although the third-order optical nonlinearity is always accompanied with the optical absorption, the figure of merit in the high frequency region is still attractive due to the presence of *weak* optical absorption. Thus, we believe that graded particles have potential applications in obtaining the optimal figure of merit, and make the composite media more realistic for practical applications.

Finally, we focus on the effect of $\gamma(r)$ on the nonlinear optical property in Fig. 5.22. As evident from the results, the variation of k_γ plays an important role in the magnitude of the effective optical properties, particularly at the surface plasmon resonance frequency.

We have developed an NDEDA (nonlinear differential effective dipole approximation) to calculate the effective linear and nonlinear dielectric responses of composite media containing nonlinear graded inclusions. The results obtained from the NDEDA are compared with the exact solutions derived from a first-principles approach for the power-law dielectric

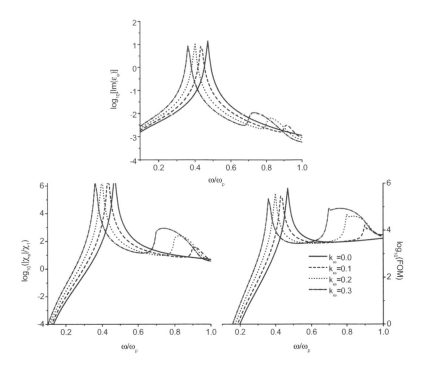

Figure 5.21. (a) The linear optical absorption $\mathrm{Im}(\varepsilon_e)$, (b) the enhancement of the third-order optical nonlinearity $|\chi_e|/\chi_1$, and (c) the figure of merit $\equiv |\chi_e|/\mathrm{Im}(\varepsilon_e)$ versus the incident angular frequency ω/ω_p for dielectric-constant gradation profile $\varepsilon(r) = 1 - \omega_p^2(r)/[\omega(\omega + i\gamma(r))]$ with $\omega_p(r) = \omega_p(1 - k_\omega \cdot r/a)$ and $\gamma(r) = 0.01\omega_p$. Parameters: $\varepsilon_2 = 1.77$ and $f = 0.05$. After Ref. [122].

gradation profiles, and the excellent agreement between them has been shown. We should remark that the exact solutions are also obtainable for the linear dielectric gradation profiles with small slopes (the derivation not shown here). In this case, the excellent agreement between the two methods can be shown as well since the NDEDA is valid indeed for arbitrary gradation profiles. In general, the exact solution is quite few in realistic composite research, and thus our NDEDA can be used as a benchmark.

In the above, the NDEDA was derived for the composite containing the nonlinear graded inclusions in a linear host. Interestingly, it can be readily generalized to the composite system where the graded inclusions and the host are both nonlinear [4]. In this situation, the effective third-order nonlinear response can be written as [68, 96],

$$
\begin{aligned}
\chi_e &= f\bar{\chi}(r=a)\left|\frac{3\varepsilon_2}{\bar{\varepsilon}(r=a)+2\varepsilon_2}\right|^2\left(\frac{3\varepsilon_2}{\bar{\varepsilon}(r=a)+2\varepsilon_2}\right)^2 + \chi_2(1-f) \\
&+ \chi_2 f\left(3\beta+\beta^*+\frac{18}{5}\beta^2+\frac{18}{5}|\beta|^2+\frac{6}{5}|\beta|^2\beta+\frac{2}{5}\beta^3+\frac{8}{5}|\beta|^2\beta^2\right),
\end{aligned} \tag{5.67}
$$

where $\beta \equiv [\bar{\varepsilon}(r=a)-\varepsilon_2]/[\bar{\varepsilon}(r=a)+2\varepsilon_2]$ and χ_2 is the third-order nonlinear susceptibility of the host medium. As a matter of fact, for this purpose, the perturbation method can also be adopted [124].

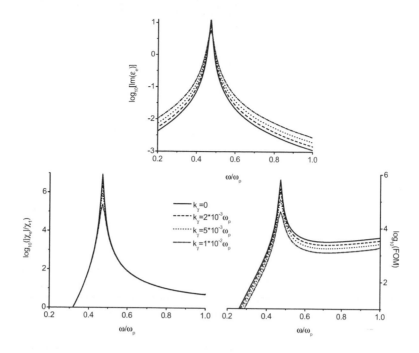

Figure 5.22. Same as Fig. 5.21, but with $\omega_p(r) = \omega_p$ and $\gamma(r) = \gamma(\infty) + k_\gamma/(r/a)$ for $\gamma(\infty) = 0.01\omega_p$. After Ref. [122].

The NDEDA is strictly valid in the dilute limit. To achieve the strong optical nonlinearity enhancement, we need possibly nonlinear inclusions with high volume fractions. In this connection, the effect of the volume fraction is expected to cause a further broadening of the resonant peak, and possibly, a desired separation of the optical absorption peak from the nonlinearity enhancement due to mutual interactions [125]. Therefore, it is of particular interest to generalize the NDEDA for treating the case of high volume fractions.

It is also instructive to develop the first-principles approach to weakly nonlinear graded composites. The perturbation approach [54] in weakly nonlinear composites is just suitable for this problem. Moreover, with the aim of the variational approach [126], the NDEDA may be applied to the cases of strong nonlinearity, where the linear part $[\varepsilon(r)]$ in Eq. (5.43) vanishes. On the other hand, based on the self-consistent mean-field approximation [90], the applicability of NDEDA to more general cases, where linear $[\varepsilon(r)]$ and nonlinear $[\chi(r)|\mathbf{E}|^2]$ parts can be comparable, may be also possible.

To sum up, we put forth an NDEDA and a first-principles approach for investigating the optical responses of nonlinear graded spherical particles. The excellent agreement between the two methods has been shown. As an application, we applied the NDEDA to discuss the surface plasmon resonance effect on the effective linear and nonlinear optical properties like the optical absorption, the optical nonlinearity enhancement, and the figure of merit. It is found that the dielectric gradation profile can be used to control the surface plasmon resonance and achieve the large FOM in the high-frequency region, where the optical absorption is quite small.

3.2. Second and Third Harmonic Generation

For optical materials with a nonlinear susceptibility other than the Kerr susceptibility, the local field can exist at all harmonic frequencies although the applied field is monochromatic. Therefore, this should be taken into account when calculating the effective nonlinear optical process. For instance, Boyd and Sipe presented a theoretical description of the nonlinear optical susceptibility for the second harmonic generation (SHG) and the Pockels effect in layered composite materials [72]. For composites containing a quadratically nonlinear phase, Levy *et al* carried out theoretical calculations on the SHG, induced third harmonic generation (THG), and optical bistability on a few simple microgeometries such as the parallel slabs, dilute spherical particles and three-component parallel slabs [127]. Later, general expressions for effective nonlinear susceptibilities for SHG and THG of composite materials were developed by Hui *et al.* in terms of the second-order and third-order nonlinear susceptibilities of the components [75, 87]. As an illustration, they applied the formulae to a dilute suspension of coated particles with the shell having a nonlinear susceptibility for second (or third) harmonic generation [77].

Below we shall investigate the effective nonlinear susceptibilities for second and third harmonic generations in nonlinear composites containing a dilute suspension of graded spherical particles. It is assumed that the graded particles have both quadratic and cubic nonlinear susceptibilities, while the host medium is linear. Based on the NDEDA (nonlinear differential effective dipole approximation), we shall derive differential equations for the equivalent nonlinear susceptibilities for SHG ($\bar{d}_{ijk}$) and THG ($\bar{\chi}_{ijkl}$). Whereafter, the composite system consisting of nonlinear graded particles can be equivalently regarded as the one containing homogeneous particles with SHG susceptibility ($\bar{d}_{ijk}$) and THG susceptibility ($\bar{\chi}_{ijkl}$) randomly embedded in a host medium. As a consequence, the effective SHG and THG susceptibilities can be obtained by using the subsistent expressions in the dilute limit. We perform numerical calculations with focus on the problem how the dielectric gradation in the particles affects the effective SHG and THG susceptibilities. On the other hand, we study alternative interesting composites, namely, the graded metallic films [14]. Due to the simple layered geometry, we achieve the effective SHG and THG susceptibilities directly. To this end, it is shown that the presence of gradation in metallic materials yields large enhancement in SHG and THG in the low-frequency region.

A. Model and definition of effective nonlinear susceptibilities for SHG and THG

We consider a macroscopically inhomogeneous material containing a dilute suspension of identical graded spherical particles with radius a and the volume fraction p in a linear host with the isotropic dielectric constant ε_2^ω. If we include both quadratic and cubic nonlinearities in the graded particles, the local constitutive relation between the displacement field **D** and the electric field **E** in the static case will be

$$D_i = \sum_{j=x}^{z} \varepsilon_{ij}(r)E_j + \sum_{j,k=x}^{z} d_{ijk}(r)E_jE_k + \sum_{j,k,l=x}^{z} \chi_{ijkl}(r)E_jE_kE_l, \quad i=x,y,z, \qquad (5.68)$$

where D_i and E_i are the ith component of **D** and **E**, and d_{ijk} (χ_{ijkl}) is the components of second-order (third-order) nonlinear susceptibility tensor of the particles. Here ε_{ij} denotes the linear dielectric function, which is assumed to be isotropic, i.e., $\varepsilon_{ij} = \varepsilon\delta_{ij}$. Moreover, for simplicity, we have supposed that the graded particles are radially inhomogeneous, and hence $\varepsilon(r)$, $d_{ijk}(r)$ and $\chi_{ijkl}(r)$ are radial functions.

When a monochromatic external field $E_0(t) = E_{0,z}(\omega)e^{-i\omega t} + c.c$ is applied, say along z-axis, in view of the existence of second-order nonlinearity inside the graded particles, the local potentials and fields will be generated at all harmonic frequencies. Thus at finite frequencies, the constitutive relation in graded particles will be [87, 127],

$$D_i = \sum_{n=-\infty}^{\infty} D_i(n\omega)e^{-in\omega t}, \qquad (5.69)$$

with

$$
\begin{aligned}
D_i(n\omega) &= \varepsilon^{(n\omega)}(r)E_i(n\omega) + \sum_{j,k=x}^{z}\sum_{m=-\infty}^{\infty} d_{ijk}^{[(n-m)\omega,m\omega]}(r)E_j[(n-m)\omega]E_k(n\omega) \\
&+ \sum_{j,k,l=x}^{z}\sum_{p,q=-\infty}^{\infty} \chi_{ijkl}^{[(n-p-q)\omega,p\omega,q\omega]}(r)E_j[(n-p-q)\omega]E_k(p\omega)E_l(q\omega),
\end{aligned}
$$

where $\varepsilon^{(n\omega)}(r)$ denotes the frequency dependent linear dielectric constant, while $d_{ijk}^{[(n-m)\omega,m\omega]}(r)$ and $\chi_{ijkl}^{[(n-p-q)\omega,p\omega,q\omega]}(r)$ represent, respectively, the second-order and third-order nonlinear susceptibilities of different harmonics.

Hui *et al* developed general expressions for the effective second and third harmonic susceptibilities of the composite media. For the present system, the effective nonlinear SHG susceptibility $d_{e,ijk}^{(\omega,\omega)}$ can be written in the form [75]

$$d_{e,ijk}^{(\omega,\omega)} = \langle K_{il}^{2\omega} d_{lmn}^{(\omega,\omega)}(r) K_{jm}^{\omega} K_{kl}^{\omega}\rangle, \qquad (5.70)$$

where $K_{il}^{\omega} \equiv E_l(\omega)/E_{0,i}(\omega)$ is the local-field factor giving the l-th Cartesian component of linear electric field inside graded particles when the external field E_0 is applied along the i-th direction at frequency ω. $\langle \cdots \rangle$ stands for the spatial average of $\cdots$. We keep the convention that repeated indices such as l, m, n in Eq.(5.70) should be summed over.

Similarly, the effective nonlinear susceptibility for general three wave mixing, which will be used in the following derivations, is given as

$$d_{e,ijk}^{(\omega_1,\omega_2)} = \langle K_{il}^{\omega_1+\omega_2} d_{lmn}^{(\omega_1,\omega_2)}(r) K_{jm}^{\omega_1} K_{kl}^{\omega_2}\rangle. \qquad (5.71)$$

For effective THG susceptibility, one yields [87]

$$\chi_{e,ijkl}^{(\omega,\omega,\omega)} = \left\langle K_{im}^{3\omega}\left\{2d_{mnp}^{(\omega,2\omega)}(r)\left[\frac{(K_{rn}^{2\omega}-\mathbf{I}_{rn})}{\delta\varepsilon^{(2\omega)}(r)}\right]d_{rst}^{(\omega,\omega)}(r) + \chi_{mstp}^{(\omega,\omega,\omega)}(r)\right\}K_{js}^{\omega}K_{kt}^{\omega}K_{lp}^{\omega}\right\rangle, \quad (5.72)$$

where $\mathbf{I}$ is a unit matrix and $\delta\varepsilon^{2\omega} \equiv \varepsilon^{2\omega}(r) - \varepsilon_e^{2\omega}$, $\varepsilon_e^{2\omega}$ being the effective linear dielectric constant at frequency 2ω. Eq. (5.72) includes two contributions to the effective THG susceptibility. One results from the intrinsic third harmonic susceptibility of the graded particles, and the other is due to the inducement of the second harmonic susceptibility of the particles.

In what follows, we shall take two steps to investigate the effective SHG and THG susceptibilities of the nonlinear graded composites. First, we develop the nonlinear differential effective dipole approximation to derive the equivalent dielectric constant $\bar{\varepsilon}^{\omega}$, equivalent

SHG (THG) susceptibility $\bar{d}_{ijk}^{(\omega,\omega)}$ ($\bar{\chi}_{ijkl}^{(\omega,\omega,\omega)}$) of the nonlinear graded particles. Then, the composite system can be regarded as the one consisting of a volume fraction p ($p \ll 1$) of homogeneous particles with linear dielectric constant $\bar{\varepsilon}^{\omega}$, SHG susceptibility $\bar{d}_{ijk}^{(\omega,\omega)}$ and THG susceptibility $\bar{\chi}_{ijkl}^{(\omega,\omega,\omega)}$ randomly embedded in a linear host medium with dielectric constant ε_2^{ω}.

B. Effective nonlinear susceptibility for second harmonic generation

We consider the composite media in which the nonlinear graded particles with linear dielectric response $\varepsilon^{\omega}(r)$ and second-order nonlinear susceptibility $d_{ijk}^{(\omega,\omega)}(r)$ are randomly embedded in a linear host medium. For the analysis of the dielectric properties of the graded particles, we resort to the differential effective dipole approximation [13, 65]. The graded particles are assumed to be constructed by adding shells gradually. In detail, we begin with an infinitesimal spherical core of the linear dielectric constant $\varepsilon^{\omega}(0)$ and second-order non-linear optical susceptibility $d_{ijk}^{(\omega,\omega)}(0)$; and keep on add shells up to $r = a$, with dielectric profiles $\varepsilon^{\omega}(r)$ and $d_{ijk}^{(\omega,\omega)}(r)$ at radius r. At r, we have an inhomogeneous spherical particle with radially inhomogeneous dielectric profiles $\varepsilon^{\omega}(r)$ and $d_{ijk}^{(\omega,\omega)}(r)$. Then, we regard the inhomogeneous sphere of radius r as a homogeneous sphere with equivalent dielectric properties $\bar{\varepsilon}^{\omega}(r)$ and $\bar{d}_{ijk}^{(\omega,\omega)}$. In this connection, both spheres should induce the same dipole moments. Further, an inhomogeneous sphere of radius $r + dr$ can be replaced by a coated one containing a spherical core of radius r with linear dielectric constant $\bar{\varepsilon}^{\omega}(r)$, and non-linear susceptibility $\bar{d}_{ijk}^{(\omega,\omega)}(r)$, and a shell of outermost radius $r + dr$ with linear dielectric constant $\varepsilon^{\omega}(r)$ and nonlinear susceptibility $d_{ijk}^{(\omega,\omega)}(r)$.

When a monochromatic external field is applied, we want to solve the potential functions in the core, the shell. The solution is [71]

$$\phi_c = -E_{0,z}(\omega)A(\omega)\rho\cos\theta \qquad \rho \leq r$$
$$\phi_s = -E_{0,z}(\omega)\left(B(\omega)\rho - \frac{C(\omega)r^3}{\rho^2}\right)\cos\theta \qquad r < \rho \leq r + dr. \qquad (5.73)$$

Here the coefficients $A(\omega)$, $B(\omega)$, and $C(\omega)$ can be determined from the appropriate boundary conditions, and we have

$$A(\omega) = \frac{9\varepsilon_2^{\omega}\varepsilon^{\omega}(r)}{Q(\omega)}, \qquad B(\omega) = \frac{3\varepsilon_2^{\omega}[\bar{\varepsilon}^{\omega}(r) + 2\varepsilon^{\omega}(r)]}{Q(\omega)},$$

$$C(\omega) = \frac{3\varepsilon_2^{\omega}[\bar{\varepsilon}^{\omega}(r) - \varepsilon^{\omega}(r)]}{Q(\omega)},$$

where $\lambda \equiv [r/(r + dr)]^3$ and

$$Q(\omega) = [\varepsilon^{\omega}(r) + 2\varepsilon_2^{\omega}][\bar{\varepsilon}^{\omega}(r) + 2\varepsilon^{\omega}(r)] + 2\lambda[\varepsilon^{\omega}(r) - \varepsilon_2^{\omega}][\bar{\varepsilon}^{\omega}(r) - \varepsilon^{\omega}(r)].$$

The equivalent linear dielectric constant for a graded particle of radius r can be readily obtained as

$$\frac{d\bar{\varepsilon}^{\omega}(r)}{dr} = \frac{[\varepsilon^{\omega}(r) - \bar{\varepsilon}^{\omega}(r)][\bar{\varepsilon}^{\omega}(r) + 2\varepsilon^{\omega}(r)]}{r\varepsilon^{\omega}(r)}. \qquad (5.74)$$

Next, we aim at deriving the equivalent second-order nonlinear susceptibility for SHG $\bar{d}_{ijk}^{(\omega,\omega)}(r)$ of the inhomogeneous graded particles. For the coated particle of radius $r+dr$, based on Eq. (5.70), one yields

$$\langle K_{il}^{2\omega}d_{lmn}^{(\omega,\omega)}K_{jm}^{\omega}K_{kn}^{\omega}\rangle_{\rho\leq r+dr} = \lambda\langle K_{il}^{2\omega}d_{lmn}^{(\omega,\omega)}K_{jm}^{\omega}K_{kn}^{\omega}\rangle_{\rho\leq r}$$
$$+(1-\lambda)\langle K_{il}^{2\omega}d_{lmn}^{(\omega,\omega)}K_{jm}^{\omega}K_{kn}^{\omega}\rangle_{r<\rho\leq r+dr} \qquad (5.75)$$

In the limit $dr \to 0$, the left-hand side of Eq. (5.75) can be expressed as

$$\langle K_{il}^{2\omega}d_{lmn}^{(\omega,\omega)}K_{jm}^{\omega}K_{kn}^{\omega}\rangle_{\rho\leq r+dr}$$
$$= \bar{d}_{ijk}^{(\omega,\omega)}(r+dr)\left(\frac{3\varepsilon_2^{\omega}}{\bar{\varepsilon}^{\omega}(r+dr)+2\varepsilon_2^{\omega}}\right)^3$$
$$= F(2\omega)F^2(\omega)\left[\bar{d}_{ijk}^{(\omega,\omega)}(r)+\frac{d\bar{d}_{ijk}^{(\omega,\omega)}(r)}{dr}dr\right.$$
$$\left.- \bar{d}_{ijk}^{(\omega,\omega)}(r)\left(\frac{d\bar{\varepsilon}^{2\omega}(r)/dr}{\bar{\varepsilon}^{2\omega}(r)+2\varepsilon_2^{2\omega}}+\frac{2d\bar{\varepsilon}^{\omega}(r)/dr}{\bar{\varepsilon}^{\omega}(r)+2\varepsilon_2^{\omega}}\right)dr\right] \qquad (5.76)$$

with $F(\omega) = 3\varepsilon_2(\omega)/[\bar{\varepsilon}^{\omega}(r)+2\varepsilon_2(\omega)]$.

The first part of the right-hand side of Eq. (5.75) has the form

$$\lambda\langle K_{il}^{2\omega}d_{lmn}^{(\omega,\omega)}K_{jm}^{\omega}K_{kn}^{\omega}\rangle_{\rho\leq r} = F(2\omega)F^2(\omega)\bar{d}_{ijk}^{(\omega,\omega)}(r)\left[1+\frac{y(2\omega)+2y(\omega)-3}{r}dr\right] \qquad (5.77)$$

with

$$y(\omega) = 2\frac{[\varepsilon^{\omega}(r)-\varepsilon_2^{\omega}][\bar{\varepsilon}^{\omega}(r)-\varepsilon^{\omega}(r)]}{\varepsilon^{\omega}(r)[\bar{\varepsilon}^{\omega}(r)+2\varepsilon_2^{\omega}]}.$$

To proceed the derivation, we make the assumption that the components of second-order nonlinear susceptibility in graded particles $d_{ijk}^{(\omega_1,\omega_2)}$ will be zero, if the indices are not equal. In other words, one has $d_{lll} \neq 0$ for $l = x,y,z$ only. However, our calculations can also be carried out even through above restrictions are released.

Then, we write the second part of the right-hand side of Eq. (5.75) as

$$(1-\lambda)\langle K_{il}^{2\omega}d_{lmn}^{(\omega,\omega)}K_{jm}^{\omega}K_{kn}^{\omega}\rangle_{r<\rho\leq r+dr}$$
$$= 3\frac{dr}{r}\sum_{l=x}^{z}\left\langle d_{lll}^{(\omega,\omega)}\frac{E_l(2\omega)}{E_{0,i}(2\omega)}\cdot\frac{E_l(\omega)}{E_{0,j}(\omega)}\cdot\frac{E_l(\omega)}{E_{0,k}(\omega)}\right\rangle_{r<\rho\leq r+dr}$$
$$\equiv 3\frac{dr}{r}G_{ijk}^{(\omega,\omega)}, \qquad (5.78)$$

where

$$\frac{E_l(\omega)}{E_{0,i}(\omega)} = \begin{cases} B(\omega)-\frac{C(\omega)r^3}{\rho^3}+\frac{3C(\omega)r^3}{\rho^5}i^2 & \text{if } i=l \\ \frac{3C(\omega)r^3}{\rho^5}il & \text{if } i\neq l, \end{cases}$$

with $\rho = x^2 + y^2 + z^2$, and

$$\langle f(x,y,z)\rangle_{r<\rho\leq r+dr} = \lim_{dr\to 0}\frac{1}{4\pi r^2 dr}\int\int\int f(x,y,z)dxdydz$$

$$= \lim_{dr\to 0}\frac{1}{4\pi r^2 dr}\int_r^{r+dr}\int_0^\pi\int_0^{2\pi} f(\rho\sin\theta\cos\phi,\rho\sin\theta\sin\phi,\rho\cos\phi)\rho^2\sin\theta d\rho d\theta d\phi.$$

The substitution of Eqs. (5.76)-(5.78) into Eq. (5.75) leads to

$$\frac{d\bar{d}_{ijk}^{(\omega,\omega)}(r)}{dr} = \bar{d}_{ijk}^{(\omega,\omega)}(r)\left(2\frac{d\bar{\varepsilon}^\omega(r)/dr}{\bar{\varepsilon}^\omega(r)+2\varepsilon_2^\omega}+\frac{d\bar{\varepsilon}^{2\omega(r)}/dr}{\bar{\varepsilon}^{2\omega}(r)+2\varepsilon_2^{2\omega}}\right.$$

$$\left.+\frac{2y(\omega)+y(2\omega)-3}{r}\right)+\frac{3G_{ijk}^{(\omega,\omega)}}{rF(2\omega)F^2(\omega)}. \tag{5.79}$$

Eqs. (5.95) and (5.79) are the main results of this section, and they provide the numerical (or analytical) solutions for the equivalent linear dielectric constant $\bar{\varepsilon}^\omega(r)$ and nonlinear susceptibility for second harmonic generation $\bar{d}_{ijk}^{(\omega,\omega)}(r)$ of graded spherical particles of radius r. These equations can be integrated as long as the gradation profiles, $\varepsilon^\omega(r)$ and $d_{ijk}^{(\omega,\omega)}(r)$, and the initial conditions are given. It should be remarked that the above derivations are valid for arbitrary gradation profiles.

It is straightforward to generalize Eq. (5.79) to study the equivalent second-order nonlinear susceptibility for general three-wave mixing $\bar{d}_{ijk}^{(\omega_1,\omega_2)}$. We have

$$\frac{d\bar{d}_{ijk}^{(\omega_1,\omega_2)}(r)}{dr} = \bar{d}_{ijk}^{(\omega_1,\omega_2)}(r)\left[\frac{d\bar{\varepsilon}^{\omega_1}(r)/dr}{\bar{\varepsilon}^{\omega_1}(r)+2\varepsilon_2^{\omega_1}}+\frac{d\bar{\varepsilon}^{\omega_2}(r)/dr}{\bar{\varepsilon}^{\omega_2}(r)+2\varepsilon_2^{\omega_2}}+\frac{d\bar{\varepsilon}^{\omega_1+\omega_2}(r)/dr}{\bar{\varepsilon}^{\omega_1+\omega_2}(r)+2\varepsilon_2^{\omega_1+\omega_2}}\right.$$

$$\left.+\frac{y(\omega_1)+y(\omega_2)+y(\omega_1+\omega_2)-3}{r}\right]$$

$$+\frac{3G_{ijk}^{(\omega_1,\omega_2)}}{rF(\omega_1)F(\omega_2)F(\omega_1+\omega_2)} \tag{5.80}$$

with

$$G_{ijk}^{(\omega_1,\omega_2)} = \sum_{l=x}^{z}\left\langle d_{lll}^{(\omega_1,\omega_2)}\frac{E_l(\omega_1+\omega_2)}{E_{0,i}(\omega_1+\omega_2)}\cdot\frac{E_l(\omega_1)}{E_{0,j}(\omega_1)}\cdot\frac{E_l(\omega_2)}{E_{0,k}(\omega_2)}\right\rangle_{r<\rho\leq r+dr}.$$

Now we are in a position to investigate the effective linear dielectric response ε_e^ω and nonlinear susceptibility for SHG $d_{e,ijk}^{(\omega,\omega)}$ of the whole system. In the dilute limit, i.e., $p\to 0$, we have [75, 127]

$$\varepsilon_e^\omega = \varepsilon_2^\omega + 3p\varepsilon_2^\omega\frac{\bar{\varepsilon}^\omega(r=a)-\varepsilon_2^\omega}{\bar{\varepsilon}^\omega(r=a)+2\varepsilon_2^\omega} \tag{5.81}$$

and

$$d_{e,ijk}^{(\omega,\omega)} = p\bar{d}_{ijk}^{(\omega,\omega)}(r=a) \left[\frac{3\varepsilon_2^{2\omega}}{\bar{\varepsilon}^{2\omega}(r=a)+2\varepsilon_2^{2\omega}} \right] \left[\frac{3\varepsilon_2^{\omega}}{\bar{\varepsilon}^{\omega}(r=a)+2\varepsilon_2^{\omega}} \right]^2. \qquad (5.82)$$

Although our formulae present all components for effective nonlinear susceptibility of SHG, here for simplicity, we only perform some numerical calculations on the z component $d_{e,zzz}^{(\omega,\omega)}$. In this case $G_{zzz}^{(\omega,\omega)}$ admits

$$\begin{aligned}
G_{zzz}^{(\omega,\omega)} &= d_{zzz}^{(\omega,\omega)}(r)\{[B_l^2(\omega)+\frac{28}{35}C_l^2(\omega)]B_l(2\omega)\\
&\quad +\frac{8}{35}[7B_l(\omega)+2C_l(\omega)]C_l(\omega)C_l(2\omega)\},
\end{aligned} \qquad (5.83)$$

where

$$\begin{aligned}
B_l(\omega) &= \lim_{dr\to 0} B(\omega) = \frac{\varepsilon_2^{\omega}[\bar{\varepsilon}^{\omega}(r)+2\varepsilon^{\omega}(r)]}{\varepsilon^{\omega}(r)[\bar{\varepsilon}^{\omega}(r)+2\varepsilon_2^{\omega}]},\\
C_l(\omega) &= \lim_{dr\to 0} C(\omega) = \frac{\varepsilon_2^{\omega}[\bar{\varepsilon}^{\omega}(r)-\varepsilon^{\omega}(r)]}{\varepsilon^{\omega}(r)[\bar{\varepsilon}^{\omega}(r)+2\varepsilon_2^{\omega}]}.
\end{aligned}$$

Note that only the component $d_{zzz}^{(\omega,\omega)}$ contributes to the effective SHG.

Substituting Eq. (5.83) into Eq. (5.79), and with Eqs. (5.95) and (5.82), we are able to investigate the possible enhancement of effective SHG susceptibility $d_e^{(\omega,\omega)}$ of graded composites in which nonlinear graded inclusions are randomly embedded in a linear dielectric host. We consider the graded spherical particles to be a Drude-like metal, which has a linear dielectric constant of the form [77]

$$\varepsilon^{\omega}(r) = 1 - \frac{\omega_p^2(r)}{\omega(\omega+i/\tau)} \qquad r \le a, \qquad (5.84)$$

where γ is the relaxation rate, and $\omega_p(r)$ represents the plasma-frequency gradation profile. For numerical calculations, $\omega_p(r)$ is assumed to be [13]

$$\omega_p(r) = \omega_p(1-k_{\omega}\frac{r}{a}), \qquad (5.85)$$

where k_{ω} is a dimensionless constant (gradient). We take $\omega_p = 2.28 \times 10^{16} s^{-1}$ and $\tau = 6.9 \times 10^{-15} s$, which correspond to these of bulk aluminum, while the host medium is assumed to have a frequency-independent dielectric constant $\varepsilon_2^{\omega} = 1.76$ (static dielectric constant of water). Furthermore, to highlight the composite effects, we set $d_{zzz}^{(\omega,\omega)}$ to be independent of both r and ω.

Figure 5.23 shows the enhancement of effective nonlinear susceptibility for SHG $d_{e,zzz}^{(\omega,\omega)}/d_{zzz}^{(\omega,\omega)}$ versus the normalized frequency ω/ω_p for various k_{ω}. For nonlinear particles without the gradation in the plasma frequency, i.e., $k_{\omega}=0$, two surface plasmon resonances appear.

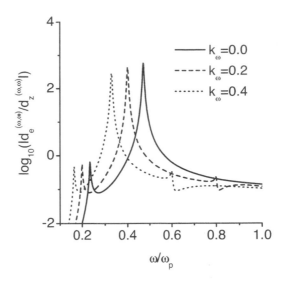

Figure 5.23. The enhancement of effective SHG susceptibility $d_e^{(\omega,\omega)}/d_z^{(\omega,\omega)} \equiv d_{e,zzz}^{(\omega,\omega)}/d_{zzz}^{(\omega,\omega)}$ versus the normalized incident angular frequency ω/ω_p for $p = 0.05$ and various k_ω. After Ref. [128].

One is located at the surface resonant frequency $\omega_c \approx \omega_p/\sqrt{1+2\varepsilon_2}$ accompanied with sharp enhancement peak in $d_{e,zzz}$, while the other arises at half of ω_c with small SHG susceptibility. Such behavior is different from that observed for effective third-order nonlinear susceptibility [114]. The difference is due to the fact that the SHG susceptibility intrinsically involves two different frequency. To one's interest, when a plasma-frequency gradation profile is taken into account, the resonant frequency exhibits red-shifted as expected. Moreover, the presence of the plasma-frequency gradation results in the broad continuous spectrum in the high-frequency region. This is in analogy to that of effective cubic nonlinear susceptibility in our previous work [13, 74].

C. *Effective nonlinear susceptibility for third harmonic generation*

In this section, we would like to derive the effective THG susceptibility of composite media consisting of a dilute suspension of graded particles with quadratic and cubic nonlinearities. To solve effective THG susceptibility is more complicated than to solve SHG susceptibility. As for THG, two effects must be taken into account. One results from the intrinsic THG susceptibility of the nonlinear graded particles, and the other is the induced THG due to the presence of SHG susceptibility of graded particles. Again, we assume that of the possible components d_{ijk} and χ_{ijk} in the graded particles, only $d_{iii}(r)$ and $\chi_{iiii}(r)$ are nonzero. Moreover, we concentrate on the z-component of the effective THG susceptibility $\chi_{e,zzzz}^{(\omega,\omega,\omega)}$.

After we perform rather bulky calculations, the equivalent THG susceptibility of the graded inclusions is found to be

$$
\frac{d\bar{\chi}_{zzzz}^{(\omega,\omega,\omega)}(r)}{dr} = \frac{2}{\bar{\varepsilon}^{2\omega}(r)+2\varepsilon_2^{2\omega}} \frac{d}{dr}\left[\bar{d}_{znz}^{(\omega,2\omega)}(r)\bar{d}_{nzz}^{(\omega,\omega)}(r)\right] - \frac{2\bar{d}_{znz}^{(\omega,2\omega)}(r)\bar{d}_{nzz}^{(\omega,\omega)}(r)}{\left[\bar{\varepsilon}^{2\omega}(r)+2\varepsilon_2^{2\omega}\right]^2}
$$

$$
\times \frac{d\bar{\varepsilon}^{2\omega}(r)}{dr} + \left[\bar{\chi}_{zzzz}^{(\omega,\omega,\omega)}(r) - \frac{2\bar{d}_{znz}^{(\omega,2\omega)}(r)\bar{d}_{nzz}^{(\omega,\omega)}(r)}{\bar{\varepsilon}^{2\omega}(r)+2\varepsilon_2^{2\omega}}\right]
$$

$$
\times \left[\frac{d\bar{\varepsilon}^{3\omega}(r)/dr}{\bar{\varepsilon}^{3\omega}(r)+2\varepsilon_2^{3\omega}} + \frac{3d\bar{\varepsilon}^{\omega}(r)/dr}{\bar{\varepsilon}^{\omega}(r)+2\varepsilon_2^{\omega}}\right]
$$

$$
+ \frac{1}{r}\left\{\left[\bar{\chi}_{zzzz}^{(\omega,\omega,\omega)}(r) - \frac{2\bar{d}_{znz}^{(\omega,2\omega)}(r)\bar{d}_{nzz}^{(\omega,\omega)}(r)}{\bar{\varepsilon}^{2\omega}(r)+2\varepsilon_2^{2\omega}}\right][y(3\omega)+3y(\omega)-3]\right.
$$

$$
\left.+ \frac{2\bar{d}_{znz}^{(\omega,2\omega)}(r)\bar{d}_{nzz}^{(\omega,\omega)}(r)F(2\omega)y(2\omega)}{\bar{\varepsilon}^{2\omega}(r)-\varepsilon_2^{2\omega}}\right\} + \frac{3H_{zzzz}}{rF(3\omega)F^3(\omega)} \qquad (5.86)
$$

where $\bar{d}_{znz}^{(\omega,2\omega)}(r)\bar{d}_{nzz}^{(\omega,\omega)}(r) \equiv \sum\limits_{n=x}^{z}[\bar{d}_{znz}^{(\omega,2\omega)}(r)\bar{d}_{nzz}^{(\omega,\omega)}(r)]$ and H_{zzzz} admits

$$
H_{zzzz} = \frac{54}{385}\frac{d_{xxx}^{(\omega,2\omega)}(r)d_{xxx}^{(\omega,\omega)}(r)+d_{yyy}^{(\omega,2\omega)}(r)d_{yyy}^{(\omega,\omega)}(r)}{\varepsilon^{2\omega}(r)-\varepsilon_2^{2\omega}}C_l^3(\omega)
$$

$$
[11B_l(2\omega)+4C_l(2\omega)]C_l(3\omega) + \frac{27}{35}\left[\chi_{xxxx}^{(\omega,\omega,\omega)}(r)+\chi_{yyyy}^{(\omega,\omega,\omega)}(r)\right.
$$

$$
\left.-2\frac{d_{xxx}^{(\omega,2\omega)}(r)d_{xxx}^{(\omega,\omega)}(r)+d_{yyy}^{(\omega,2\omega)}(r)d_{yyy}^{(\omega,\omega)}(r)}{\varepsilon^{2\omega}(r)-\varepsilon_2^{2\omega}}\right]C_l^3(\omega)C_l(3\omega)
$$

$$
+ \frac{2}{385}\frac{d_{zzz}^{(\omega,2\omega)}(r)d_{zzz}^{(\omega,\omega)}(r)}{\varepsilon^{2\omega}(r)-\varepsilon_2^{2\omega}}\left[385B_l^3(\omega)B_l(2\omega)B_l(3\omega)\right.
$$

$$
+924B_l(\omega)B_l(2\omega)B_l(3\omega)C_l^2(\omega)+176B_l(2\omega)B_l(3\omega)C_l^3(\omega)
$$

$$
+924B_l^2(\omega)B_l(3\omega)C_l(\omega)C_l(2\omega)+528B_l(\omega)B_l(3\omega)C_l^2(\omega)C_l(2\omega)
$$

$$
+528B(3\omega)C_l^3(\omega)C_l(2\omega)+924B_l^2(\omega)B_l(2\omega)C_l(\omega)C_l(3\omega)
$$

$$
+528B_l(\omega)B_l(2\omega)C_l^2(\omega)C_l(3\omega)+528B_l(2\omega)C_l^3(\omega)C_l(3\omega)
$$

$$
+308B_l^3(\omega)C_l(2\omega)C_l(3\omega)+528B_l^2(\omega)C_l(\omega)C_l(2\omega)C_l(3\omega)
$$

$$
\left.+1584B_l(\omega)C_l^2(\omega)C_l(2\omega)C_l(3\omega)+640C_l^3(\omega)C_l(2\omega)C_l(3\omega)\right]
$$

$$
+ \frac{1}{35}\left[\chi_{zzzz}^{(\omega,\omega,\omega)}(r) - 2\frac{d_{zzz}^{(\omega,2\omega)}(r)d_{zzz}^{(\omega,\omega)}(r)}{\varepsilon^{2\omega}-\varepsilon_2^{2\omega}}\right][35B_l^3(\omega)B_l(3\omega)
$$

$$
+84B_l^2(\omega)C_l(\omega)C_l(3\omega)+16C_l^3(\omega)B_l(3\omega)+48C_l^3(\omega)C_l(3\omega)
$$

$$
+84B_l(\omega)C_l^2(\omega)B_l(3\omega)+48B_l(\omega)C_l^2(\omega)C_l(3\omega)]. \qquad (5.87)
$$

The effective THG susceptibility of the whole composite is then given by [87]

$$\chi_{e,zzzz}^{(\omega,\omega,\omega)} = p \left[\frac{3\varepsilon_2^{3\omega}}{\bar{\varepsilon}^{3\omega}(r=a) + 2\varepsilon_2^{3\omega}} \right] \left[\frac{3\varepsilon_2^{\omega}}{\bar{\varepsilon}^{\omega}(r=a) + 2\varepsilon_2^{\omega}} \right]^3 \cdot$$
$$\left\{ \bar{\chi}_{zzzz}^{(\omega,\omega,\omega)}(r=a) - \frac{2\sum\limits_{n=x}^{z} \left[\bar{d}_{znz}^{(\omega,2\omega)}(r=a)\bar{d}_{nzz}^{(\omega,\omega)}(r=a) \right]}{\bar{\varepsilon}^{2\omega}(r=a) + 2\varepsilon_2^{2\omega}} \right\} . \qquad (5.88)$$

In Eq. (5.88), because the equivalent nonlinear susceptibility $\bar{d}_{znz}^{(\omega,2\omega)}(r=a)$ [or $\bar{d}_{nzz}^{(\omega,\omega)}(r=a)$] is dependent on the second nonlinear susceptibility $d_{iii}(r)$ of the graded particles [see Eqs. (5.79) and (5.80)], and the equivalent THG susceptibility $\bar{\chi}_{zzzz}^{(\omega,\omega,\omega)}(r=a)$ relies on both $d_{iii}^{(\omega,\omega)}(r)$ and $\chi_{iiii}^{(\omega,\omega,\omega)}(r)$ of the graded particles [see Eq. (5.97)], the effective THG susceptibility of the whole system include two terms. One is the THG susceptibility due to the intrinsic THG susceptibility in the graded particles, and the other is the induced THG susceptibility due to the presence of second-order nonlinear susceptibility in the particles. Therefore, in what follows, we discuss these two effects respectively.

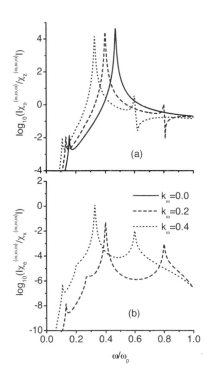

Figure 5.24. The enhancement of effective THG susceptibility $\chi_e^{(\omega,\omega,\omega)} \equiv \chi_{e,zzzz}^{(\omega,\omega,\omega)}$ due to the presence of intrinsic THG susceptibility of graded particles. (a) $\chi_z^{(\omega,\omega,\omega)} \equiv \chi_{zzzz}^{(\omega,\omega,\omega)}$ being the only nonzero component; (b) $\chi_x^{(\omega,\omega,\omega)} \equiv \chi_{xxxx}^{(\omega,\omega,\omega)}$ being the only nonzero component. Note that $\chi_{e,zzzz}^{(\omega,\omega,\omega)}/\chi_{xxxx}^{(\omega,\omega,\omega)} = 0$ for $k_\omega = 0$. After Ref. [128].

First, we study the enhancement of the effective THG susceptibility $\chi_{e,zzzz}^{(\omega,\omega,\omega)}$ due to the presence of intrinsic THG susceptibility $\chi_{iiii}^{(\omega,\omega,\omega)}$ ($i = x, y, z$) of nonlinear graded particles. This is shown in Fig. 5.24. For nonzero $\chi_{zzzz}^{(\omega,\omega,\omega)}$, the enhancement of the effective THG susceptibility [see Fig. 5.24(a)] exhibits quite similar behavior as that of SHG susceptibility (see Fig. 5.23). However, since effective THG susceptibility is driven by three power of the local field at frequency ω, while effective SHG susceptibility is driven by two power, the enhancement factor of THG at the surface resonant frequency ω_c is found to be higher than that of SHG. Moreover, in the low-frequency region, the additional resonance with small magnitude occurs at $\omega_c/3$ instead of $\omega_c/2$ for SHG. Again, the continuous band appears if the gradation in the plasma frequency is taken into account.

It is known that for the homogeneous particles, i.e., $k_\omega = 0$, the spatial local field inside the particles will be uniform with it's direction along the direction of the applied field, say z-axis. Therefore, in this case, the THG component $\chi_{xxxx}^{(\omega,\omega,\omega)}$ ($\chi_{yyyy}^{(\omega,\omega,\omega)}$) does not contribute to the effective THG susceptibility $\chi_{e,zzzz}^{(\omega,\omega,\omega)}$. At the same time, in the case of dielectric gradation ($k_\omega \neq 0$), the local field will become spatial-dependent and possess the component E_x (E_y) even though the direction of the applied field is along z-axis. In this connection, the nonlinear THG component $\chi_{xxxx}^{(\omega,\omega,\omega)}$ ($\chi_{yyyy}^{(\omega,\omega,\omega)}$) will result in nonzero effective THG susceptibility [see Fig. 5.24(b)]. It should be noted that the enhancement factor due to the χ_{zzzz} is five orders of magnitude larger than the one due to χ_{xxxx} (χ_{yyyy}).

The induced effective THG susceptibility in the presence of the second nonlinear susceptibility as a function of ω/ω_p is shown in Fig. 5.25. In such a case, the nonlinear optical process involves the effect of first forming a 2ω component and then combining the 2ω component with a ω component to give a 3ω component. As a result, there exist three sharp peaks with predominant one located at the resonant frequency ω_c. Otherwise, the behavior shown in Fig. 5.25 is quite similar to that in Fig. 5.24. Here, we would like to mention that the enhancements of the effective THG susceptibility due to the intrinsic THG susceptibility of the graded particles and due to the inducement of second order nonlinear susceptibility have the same magnitude. Therefore, to calculate the effective THG susceptibility of the composites, one should taken into account both contributions simultaneously.

D. Effective SHG and THG susceptibilities in nonlinear graded films

In above sections, we develop the nonlinear differential effective dipole approximations for the effective SHG and THG susceptibilities of graded composite media in which graded metallic particles with weak nonlinearity are randomly embedded in the linear dielectric host. In the following, we shall consider an alternative model, namely a layered film [4] but containing nonlinear graded metal [14].

The film is formed of graded metallic material with width L. The gradation under consideration is in the direction perpendicular to the film (say along z-axis). At position z, the material possesses the linear dielectric constant $\varepsilon^\omega(z)$ with the Drude form,

$$\varepsilon^\omega(z) = 1 - \frac{\omega_p^2(z)}{\omega(\omega + i/\tau)}, \quad 0 \leq z \leq L. \tag{5.89}$$

Similarly, we adopt the plasma-frequency gradation profile

$$\omega_p(z) = \omega_p(1 - k_\omega \frac{z}{L}). \tag{5.90}$$

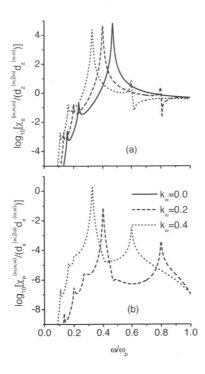

Figure 5.25. The enhancement of effective THG susceptibility $\chi_e^{(\omega,\omega,\omega)} \equiv \chi_{e,zzzz}^{(\omega,\omega,\omega)}$ due to the inducement of the second-order nonlinear susceptibility of graded particles. (a) $d_z^{(\omega,2\omega)} d_z^{(\omega,\omega)}$ being the only nonzero component; (b) $d_x^{(\omega,2\omega)} d_x^{(\omega,\omega)}$ being the only nonzero component. Note that $\chi_{e,zzzz}^{(\omega,\omega,\omega)}/(d_x^{(\omega,2\omega)} d_x^{(\omega,\omega)}) = 0$ for $k_\omega = 0$. After Ref. [128].

It has been pointed out that the position-dependent plasma frequency described by Eq. (5.90) may be realized by imposing a temperature gradient in the direction perpendicular to the film [86]. We make a further assumption that the material possesses z-independent second order nonlinear susceptibility $d_{ijk}^{(\omega_1,\omega_2)}$ and THG susceptibility $\chi_{ijkl}^{(\omega,\omega,\omega)}$. For such a simple geometry, the effective SHG and THG susceptibilities of the film can be exactly derived.

Based on Eqs. (5.70) and (5.72), the effective nonlinear susceptibilities for SHG and THG involving z-polarized light are, respectively, written as

$$d_{e,zzz}^{(\omega,\omega)} = d_{zzz}^{(\omega,\omega)} \frac{1}{L} \int_0^L \frac{\varepsilon_{e,zz}^{2\omega}}{\varepsilon^{2\omega}(z)} \cdot \left[\frac{\varepsilon_{e,zz}^{\omega}}{\varepsilon^{\omega}(z)}\right]^2 dz, \tag{5.91}$$

and

$$\chi_{e,zzzz}^{(\omega,\omega,\omega)} = -2d_{zzz}^{(\omega,2\omega)} d_{zzz}^{(\omega,\omega)} \frac{1}{L} \int_0^L \frac{\varepsilon_{e,zz}^{3\omega}}{\varepsilon^{3\omega}(z)} \cdot \left[\frac{\varepsilon_{e,zz}^{\omega}}{\varepsilon^{\omega}(z)}\right]^3 \cdot \left[\frac{1}{\varepsilon^{2\omega}(z)}\right] dz$$

$$+ \chi_{zzzz}^{(\omega,\omega,\omega)} \frac{1}{L} \int_0^L \frac{\varepsilon_{e,zz}^{3\omega}}{\varepsilon^{3\omega}(z)} \cdot \left[\frac{\varepsilon_{e,zz}^{\omega}}{\varepsilon^{\omega}(z)}\right]^3 dz, \tag{5.92}$$

where the effective linear dielectric constant $\varepsilon^{\omega}_{e,zz}$ is given by

$$\frac{1}{\varepsilon^{\omega}_{e,zz}} = \frac{1}{L} \int_0^L \frac{dz}{\varepsilon^{\omega}(z)}. \tag{5.93}$$

In deriving Eqs. (5.91) and (5.92), we have made use of the continuity of electric displacement, i.e., $\varepsilon^{\omega}(z)E_z(\omega) = \varepsilon^{\omega}_{e,zz}E_{0,z}(\omega)$.

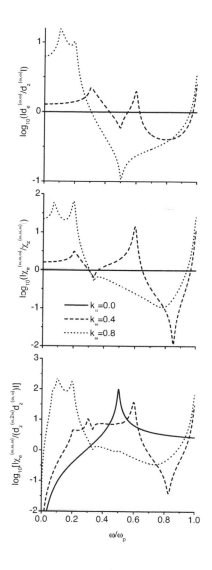

Figure 5.26. For metallic graded films, $d_e^{(\omega,\omega)}$, $\chi_e^{(\omega,\omega,\omega)}$ due to the presence of $\chi_z^{(\omega,\omega,\omega)}$, and the induced $\chi_e^{(\omega,\omega,\omega)}$ due to the second-order nonlinear susceptibility as a function of ω/ω_p for $\omega_p\tau = 50$. After Ref. [128].

In Fig. 5.26, we plot the enhancement of effective nonlinear susceptibilities for SHG and THG as a function of ω/ω_p for various k_ω. When the gradation is not included, no enhancement of effective SHG susceptibility (or THG susceptibility due to the intrinsic THG susceptibility of the metallic material) is found, as expected. In addition, the induced effective THG susceptibility in the presence of the second order nonlinear susceptibility of the component achieves large enhancement at $\omega_c = \omega_p/2$. Actually, this sharp enhancement results from the fact that at such a frequency, metallic material has a highly dispersive but predominantly real linear dielectric constant $\varepsilon^{2\omega}$. To one's interest, owing to gradation, large enhancement of effective SHG and THG are evidently observed in the low-frequency region. Therefore, one would expect that the higher enhancement of effective SHG and THG susceptibilities of graded metallic films may be achieved with the presence of sufficiently large gradient. In this connection, the graded metallic films may serve as a novel optical material for producing large enhancements in both SHG and THG signals.

With the aid of nonlinear differential effective dipole approximation (NDEDA), we have investigated the effective nonlinear susceptibilities for second harmonic generation and third harmonic generation of nonlinear composite media containing graded spherical particles. For effective THG susceptibility, we consider the contributions not only from the presence of the intrinsic THG susceptibility of the graded particles, but also from the inducement of the SHG susceptibility in the graded particles. It is found that the presence of gradation yields a continuous band in the high-frequency region and the red-shift of the dominant resonant frequency. Furthermore, we examine the composite film of grade metallic material. In such a case, we present the effective SHG and THG susceptibilities analytically. we numerically demonstrate that in the low-frequency region, large enhancement of effective SHG and THG susceptibilities is indeed achieved for sufficiently large gradient. This opens up the possibility of using graded metallic film as a suitable candidate material in improved SHG and THG devices.

The NDEDAs for the effective nonlinear susceptibilities for SHG and THG are strictly valid in the dilute limit. For nondilute grade particles, the electrostatic interactions between the graded particles should be taken into account. In this regard, the external applied field $E_{0,z}(\omega)$ in Eq. (5.109) should be replaced by the Lorentz local field, and the effective nonlinear susceptibilities can be estimated using the Clausuis-Mossotti approximation. On the other hand, our consideration is restricted in the weak nonlinearity. When the applied field is large, intrinsic optical bistability will also appear in such kind of composites containing graded particles with a quadratical nonlinearity. It has been shown that the condition under which the optical bistability appears in some simple microgeometries like parallel slabs microstructure, is quite different from those in the same microgeometries of cubic nonlinearity [127]. Therefore, it is of great interest to examine the optical bistability in the nonlinear composites containing graded particles with quadratical and cubic nonlinearities.

3.3. Nondegenerate Four-Wave Mixing

For degenerate four-wave mixing cases, the effective degenerate third-order nonlinear optical susceptibility is $\chi_e(\omega) \equiv \chi_e(\omega : -\omega, \omega, \omega)$. However, in nondegenerate four-wave mixing cases, one should impose the two pump fields with ω_1 of high intensity to generate the desired nonlinearity, while the probe field at ω_2 of lower intensity is measured. For instance, the differential absorption spectra [129] are related to the effective nondegenerate third-order nonlinear optical susceptibility (NDTNOS) $\chi_e(\omega_2) \equiv \chi_e(\omega_2 : -\omega_1, \omega_1, \omega_2)$

defined as

$$\chi_e(\omega_2)|\mathbf{E}_{0,\omega_1}|^2\mathbf{E}_{0,\omega_2}^2 = \frac{1}{V}\int \chi_i|\mathbf{E}_{\text{loc},\omega_1}|^2\mathbf{E}_{\text{loc},\omega_2}^2 dV,$$

where χ_i stands for the NDTNOS of the component i, and $\mathbf{E}_{\text{loc},\omega}$ represents the linear local field when the external field of ω $(\mathbf{E}_{0,\omega})$ is applied. This susceptibility is qualitatively different from the degenerate one in that it intrinsically involves two different frequencies. In this report, we shall generalize the NDEDA [13], which is valid for the degenerate third-order nonlinear optical susceptibility of graded composites in the dilute limit, to treat the effective NDTNOS of composite media containing spherical graded particles with high volume factions. For this purpose, the Bruggeman effective medium approximation (EMA) will be adopted. Furthermore, we also apply the NDEDA to study the effective NDTNOS of the composite with H-S microstructure.

Let us consider a two-phase composite material, in which the graded metallic particles with volume fraction p, and the dielectric grains of the dielectric constant $\varepsilon_2(\omega)$ with $1-p$ are randomly distributed. In the system, the graded particles with the same radius a are assumed to possess the weakly nonlinear displacement (**D**)-field (**E**) relation of the form, $\mathbf{D}_{1,\omega_2} = \varepsilon(r,\omega_2)\mathbf{E}_{1,\omega_2} + \chi(r,\omega_2)|\mathbf{E}_{1,\omega_1}|^2\mathbf{E}_{1,\omega_2}$, where $\varepsilon(r,\omega)$ and $\chi(r,\omega)$, respectively, stand for the linear dielectric constant and the NDTNOS of graded particles at the frequency ω. It is worth noting that both $\varepsilon(r,\omega)$ and $\chi(r,\omega)$ are radial functions.

Within the quasi-static approximation, the whole inhomogeneous composites behave as an effective homogeneous one with the effective linear dielectric constant $\varepsilon_e(\omega_2)$ and NDTNOS $\chi_e(\omega_2)$, given by

$$\langle \mathbf{D}_{\omega_2}\rangle = \varepsilon_e(\omega_2)\mathbf{E}_{0,\omega_2} + \chi_e(\omega_2)|\mathbf{E}_{0,\omega_1}|^2\mathbf{E}_{0,\omega_2}, \tag{5.94}$$

where $\langle\cdots\rangle$ denotes the spatial average.

To obtain $\varepsilon_e(\omega)$ and $\chi_e(\omega)$, we consider the composites in which both the nonlinear graded spherical inclusions and the dielectric grains are embedded in the host medium with undetermined linear dielectric constant $\varepsilon_e(\omega)$. The equivalent dielectric constant $\bar{\varepsilon}(r,\omega)$ of the graded particles at radius r receives the form [125]

$$\frac{d\bar{\varepsilon}(r,\omega)}{dr} = \frac{[\varepsilon(r,\omega)-\bar{\varepsilon}(r,\omega)][\bar{\varepsilon}(r,\omega)+2\varepsilon(r,\omega)]}{r\varepsilon(r,\omega)}. \tag{5.95}$$

Then, the effective linear dielectric constant $\varepsilon_e(\omega)$ of the whole system is self-consistently given by the Bruggeman EMA:

$$p\frac{\bar{\varepsilon}(a,\omega)-\varepsilon_e(\omega)}{\bar{\varepsilon}(a,\omega)+2\varepsilon_e(\omega)} + (1-p)\frac{\varepsilon_2(\omega)-\varepsilon_e(\omega)}{\varepsilon_2(\omega)+2\varepsilon_e(\omega)} = 0. \tag{5.96}$$

For the equivalent NDTNOS $\bar{\chi}(r,\omega_2)$ of the graded particles, we have

$$\begin{aligned}
\frac{d\bar{\chi}(r,\omega_2)}{dr} &= \bar{\chi}(r,\omega_2)\left[\sum_{n=1}^{2}\frac{nd\bar{\varepsilon}(r,\omega_n)/dr}{\bar{\varepsilon}(r,\omega_n)+2\varepsilon_e(\omega_n)} + \left(\frac{d\bar{\varepsilon}(r,\omega_1)/dr}{\bar{\varepsilon}(r,\omega_1)+2\varepsilon_e(\omega_1)}\right)^*\right] \\
&\quad + \frac{\bar{\chi}(r,\omega_2)}{r}[y(r,\omega_1)+y^*(r,\omega_1)+2y(r,\omega_2)-3] \\
&\quad + \frac{3\chi(r,\omega_2)G}{5r},
\end{aligned} \tag{5.97}$$

where

$$y(\omega) = 2\frac{[\varepsilon(r,\omega) - \varepsilon_e(\omega)][\bar{\varepsilon}(r,\omega) - \varepsilon(r,\omega)]}{\varepsilon(r,\omega)[\bar{\varepsilon}(r,\omega) + 2\varepsilon_e(\omega)]},$$

and

$$\begin{aligned}
G = \; & 5|B(\omega_1)|^2[B^2(\omega_2) + 2C^2(\omega_2)] \\
& + 4B^*(\omega_1)C(\omega_1)C(\omega_2)[2B(\omega_2) + C(\omega_2)] \\
& + 8B(\omega_2)C^*(\omega_1)C(\omega_2)[B(\omega_1) + C(\omega_1)] \\
& + 4C^*(\omega_1)C^2(\omega_2)[B(\omega_1) + 6C(\omega_1)] \\
& + 10B^2(\omega_2)|C(\omega_1)|^2,
\end{aligned}$$

with $B(\omega)[C(\omega)] = [\bar{\varepsilon}(r,\omega) \pm 2\varepsilon(r,\omega)]/[3\varepsilon(r,\omega)]$.

Next, we resort to the spectral represent theory [44] and adopt the decoupling approximation to investigate the effective NDTNOS of the composites. That is,

$$\begin{aligned}
\chi_e(\omega_2) = \; & p\bar{\chi}(a,\omega_2)\int_0^1\left|\frac{s(\omega_1)}{s(\omega_1) - x}\right|^2 m(x)\mathrm{d}x \\
& \times \int_0^1\left(\frac{s(\omega_2)}{s(\omega_2) - x}\right)^2 m(x)\mathrm{d}x,
\end{aligned} \tag{5.98}$$

where $s(\omega) \equiv \varepsilon_e(\omega)/[\varepsilon_e(\omega) - \bar{\varepsilon}(\omega)]$ and $m(x)$ is the spectral density function given in Ref. [44]. Here we mention that the decoupling procedure will be accurate when the local field is nearly uniform and less accurate when the field fluctuations are large as in a random composite near the percolation threshold.

As a model system, we consider the graded spherical particles to be a Drude-like metal, which has a linear dielectric constant of the form [13]

$$\varepsilon(r,\omega) = 1 - \frac{\omega_p^2(r)}{\omega(\omega + i\gamma)} \qquad r \le a, \tag{5.99}$$

where γ is the relaxation rate, and $\omega_p(r)$ represents the plasma-frequency gradation profile. For numerical calculations, $\omega_p(r)$ is assumed to have the form $\omega_p(r) = \omega_p(1 - k_\omega r/a)$ [13]. Furthermore, to highlight the composite effect, we set $\chi(r,\omega) \equiv \chi_1$ to be independent of both r and ω.

Fig. 5.27 displays the linear optical absorption coefficient $\alpha \sim \omega_2/\omega_p\mathrm{Im}[\sqrt{\varepsilon_e(\omega_2)}]$ versus the normalized frequency ω_2/ω_p, for $\omega_1 = \omega_p/\sqrt{1 + 2\varepsilon_2}$. Due to the electromagnetic interaction between the individual grains, there are surface plasmon resonant bands in the whole frequency region $0 < \omega < \omega_p$. Moreover, at high volume fraction which is larger than the percolation threshold $p_c = 1/3$, a Drude peak appears, characterized by a fast increase of linear absorption near $\omega \sim 0$. To one's interest, when a plasma-frequency gradation profile is taken into account, the surface resonant bands are split into two parts: one is due to randomness; the other (within the high-frequency region) results from the plasmon-frequency gradation. In particular, at high volume fraction, the presence of gradation leads to a significant decrease in the magnitude of the optical absorption band.

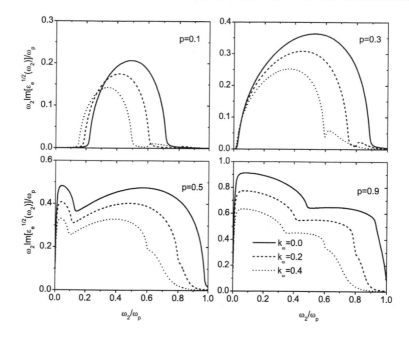

Figure 5.27. The linear optical absorption $\omega_2 \mathrm{Im}[\sqrt{\varepsilon_e(\omega_2)}]/\omega_p$ versus the normalized incident angular frequency ω_2/ω_p , for various k_ω and p. Parameter: $\omega_1 = \omega_p/\sqrt{1+2\varepsilon_2}$ (the resonant frequency of a single metallic particle in 3D). After Ref. [130].

In Fig. 5.28, we study the enhancement of the effective NDTNOS $|\chi_e(\omega_2)/\chi_1|$. Owing to gradation, the resonant bands due to randomness are caused to be red-shifted, while the other enhancement bands are induced to appear in the high-frequency region. The latter enhancement can be well understood if we regard the graded particles as a limit of multi-shells [74]. Furthermore, for high volume fraction, the enhancement of the effective NDTNOS for graded composites is larger than the one for the non-graded composites. Therefore, for high p, by choosing an appropriate gradation profile, it is possible to achieve a more prominent enhancement of the effective NDTNOS accompanied with a small linear optical absorption.

For practical applications, the most useful parameter is the FOM, defined as $|\chi_e(\omega_2)/\chi_1|/\alpha$, see Fig. 5.29. At large volume fraction, the FOM for the graded composites is apparently larger than the one for the non-graded composites.

In what follows, we shall derive the NDEDA for the Hashin-Shtrikman (HS) microgeometry. Now, we have a two-phase composite consisting of entirely coated spheres with a nonlinear graded core of dielectric properties $\varepsilon(r,\omega)$ and $\chi(r,\omega)$, and a concentric shell of $\varepsilon_2(\omega)$. For this kind of microstructure, the equivalent linear dielectric constant $\bar{\varepsilon}(r,\omega)$ of graded inclusions can still be described by Eq. (5.95). However, since the nonlinear graded particles (cores) are always surrounded by the host medium of the dielectric constant $\varepsilon_2(\omega)$, the corresponding equivalent NDTNOS can be obtained from Eq. (5.97) with $\varepsilon_e(\omega)$ being replaced by $\varepsilon_2(\omega)$.

Once $\bar{\varepsilon}(r,\omega)$ and $\bar{\chi}(r,\omega)$ are calculated, the effective linear dielectric constant of the graded composites with the HS microgeometry is described by Maxwell-Garnett approxi-

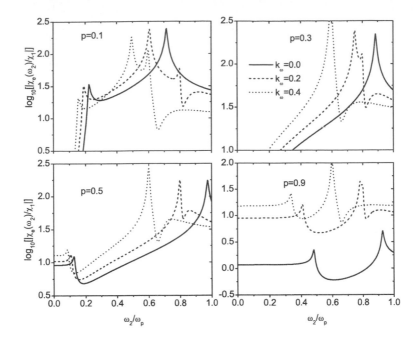

Figure 5.28.   Same as Fig. 5.27, but for the effective NDTNOS $|\chi_e(\omega_2)/\chi_1|$.   After Ref. [130].

mation (MGA),

$$\frac{\varepsilon_e(\omega_2)}{\varepsilon_2(\omega_2)} = 1 + \frac{3p[\bar{\varepsilon}(a,\omega_2) - \varepsilon_2(\omega)]}{(1-p)\bar{\varepsilon}(a,\omega_2) + (2+p)\varepsilon_2(\omega_2)}. \qquad (5.100)$$

The effective NDTNOS for H-S microgeometry is given by

$$\chi_e(\omega_2) = p\bar{\chi}(a,\omega_2)\left|\frac{3\varepsilon_2(\omega_1)}{(1-p)\bar{\varepsilon}(a,\omega_1) + (2+p)\varepsilon_2(\omega_1)}\right|^2$$

$$\times \left(\frac{3\varepsilon_2(\omega_2)}{(1-p)\bar{\varepsilon}(a,\omega_2) + (2+p)\varepsilon_2(\omega_2)}\right)^2. \qquad (5.101)$$

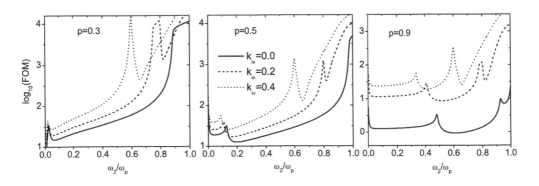

Figure 5.29. Same as Fig. 5.27, but for the FOM. After Ref. [130].

In Fig. 5.30, we display the linear absorption $\alpha \sim \omega_2/\omega_p \mathrm{Im}[\sqrt{\varepsilon_e(\omega_2)}]$ and nonlinear absorption $\beta \sim |\mathrm{Im}[\chi_e(\omega_2)/\chi_1]|$ as a function of p. Both the linear and nonlinear absorption are in direct proportion to p at low volume fraction, predicted from either the NDEDA accompanied with the EMA, or the NDEDA accompanied with the MGA. Then, the linear absorption deviates from the linear dependence on p and undergoes a sharp increase at a certain high volume fraction, dependent on the gradient k_ω. At small gradients, after the linear dependence, the nonlinear absorption for the NDEDA accompanied with the MGA goes through a maximum, and then decreases monotonically with p. For large gradient $k_\omega = 0.4$, a sharp valley appears at $p \approx 0.44$, and $\mathrm{Im}[\chi_e(\omega_2)/\chi_1]$ crossovers from the negative value to positive one. However, for the NDEDA accompanied with the EMA, broad nonlinear absorption bands are observed again. Moreover, for large gradient $k_\omega = 0.4$, a linear enhancement of nonlinear absorption is found for $0 < p < 3 \times 10^{-2}$. At volume fraction $3 \times 10^{-2} < p < 0.112$, the degree of enhancement is higher. After that, the monotonic decrease of nonlinear absorption with p comes to appear due to the formation of large clusters. All these properties are in qualitatively agreement with the experimental report [129]. In order to compare with experimental results quantitatively, we should apply the Shalaev-Sarychev theory [10, 73] by taking into account the mutual interaction effects exactly. On the other hand, since graded films can be fabricated easily, we suggest experiments be done to examine the gradation effect in the graded metallic films [14].

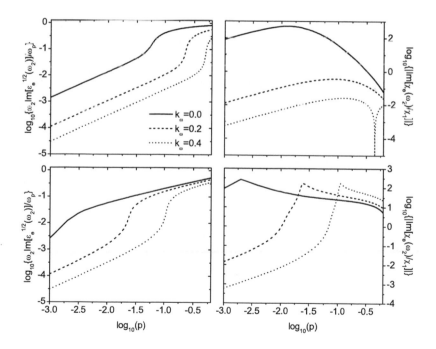

Figure 5.30. The linear and nonlinear optical absorption as a function of p, for various k_ω. The results of both the NDEDA accompanied with the MGA (upper panels) and the EMA (lower panels) are shown. After Ref. [130].

In conclusion, we have developed the NDEDA accompanied with the EMA for calculating the NDTNOS of a graded composite in which the nonlinear graded metallic particles
and the linear dielectric grains are randomly distributed. At high volume fraction, the presence of gradation was found to be helpful to achieve a large enhancement of the NDTNOS
and FOM. The effect of composite topology has also been studied by using the NDEDA
accompanied with the MGA.

4. Composites of Graded Particles...

The aforementioned gradation models were built up under the assumption that the graded
inclusions exhibit isotropic dielectric response. However, dielectric anisotropy occurs naturally due to the presence of gradation inside the particles. Moreover, there are many
inhomogeneous materials with spatial anisotropy, like polycrystal aggregates of a single
anisotropic component [131], liquid crystal droplets [132], and cell membranes containing
mobile charges [133]. In these situations, the local dielectric coefficient should be tensorial.
Thus, for a better understanding of the dielectric-anisotropy effect, it is necessary to generalize our previous isotropic gradation models [12, 63, 64, 65] accordingly. The problem
is further complicated by the fact that, in realistic composites, besides the inhomogeneity
and anisotropy, the nonlinearity plays an important role in determining the effective material properties of realistic composite media [10, 70]. In fact, the nonlinearity is a common
phenomena in realistic graded materials. And the spatial anisotropy effect has not yet been
investigated in the traditional theories. In the following, we shall develop a new theory, in
an attempt to study the effective linear and nonlinear optical properties of composite media, by taking into account the dielectric anisotropy of the nonlinear graded particles. For
the dielectric tensor of these graded inclusions, the components of the tensorial dielectric
constant of interest will be assumed to vary along the radius of the particles continuously.

*A. Model and definition of effective linear dielectric constant and third-order nonlinear
susceptibility*

Let us consider a dilute composite material, where the identical spherical inclusions
having a dielectric constant tensor $\bar{\bar{\varepsilon}}_1$, with radius a, are randomly embedded in a linear
isotropic host with (scalar) dielectric constant ε_2. Inside the anisotropic inclusions, the local constitutive relation between the displacement (**D**) and the electric field (**E**) is given
by [134]

$$D_i = \sum_j \varepsilon_{ij} E_j + \sum_{jkl} \chi_{ijkl} E_j E_k E_l^*. \qquad (5.102)$$

Here D_i and E_i are respectively the i-th Cartesian components of **D** and **E**. It is worth
remarking that ε_{ij} and χ_{ijkl} are the second-rank and fourth-rank Cartesian tensors, respectively. Throughout the section, our analysis will be limited to the case of weak nonlinearity.
In other words, the nonlinear part in Eq. (5.102) will be assumed to be small when compared
with the linear part.

In what follows, the dielectric tensor for the the anisotropic spherical inclusions is assumed to be diagonal in spherical coordinates, with a value $\varepsilon_{1t}(r)$ in the tangential directions
and $\varepsilon_{1r}(r)$ in the radial direction [18, 133, 135]. Here, both dielectric gradation profiles
$\varepsilon_{1r}(r)$ and $\varepsilon_{1t}(r)$ will be mathematically represented as radial functions [18]. In view of the

spherical symmetry, we can express the dielectric constant tensor $\bar{\varepsilon}_1(r)$ of graded particles in the form

$$\bar{\varepsilon}_1(r) = \begin{pmatrix} \varepsilon_{1r}(r) & 0 & 0 \\ 0 & \varepsilon_{1t}(r) & 0 \\ 0 & 0 & \varepsilon_{1t}(r) \end{pmatrix}. \tag{5.103}$$

Note the above form is in spherical coordinates, rather than in Cartesian coordinates. Nevertheless, it can also be represented in Cartesian coordinates by a transformation using appropriate rotation matrices.

As the graded inclusions with dielectric anisotropy are randomly oriented, the whole sample should be macroscopically isotropic. Thus, we can define the effective linear dielectric constant ε_e and the third-order nonlinear susceptibility χ_e of the whole composite as [66, 134]

$$\mathbf{D} = \varepsilon_e \mathbf{E}_0 + \chi_e |\mathbf{E}_0|^2 \mathbf{E}_0, \tag{5.104}$$

where $\langle \cdots \rangle$ stands for the spatial average, and $\mathbf{E}_0 = E_0 \mathbf{e}_z$ denotes the external applied field along z-axis. In Eq. (5.104), the effective linear dielectric constant ε_e is given by

$$\varepsilon_e \mathbf{E}_0 = \frac{1}{V} \int_V \bar{\varepsilon} \cdot \mathbf{E}_{\text{lin}} dV = f \langle \bar{\varepsilon}_1 \cdot \mathbf{E}_{1,\text{lin}} \rangle + (1-f) \varepsilon_2 \langle \mathbf{E}_{2,\text{lin}} \rangle, \tag{5.105}$$

where f is the volume fraction of the graded inclusions. Here, the subscript 'lin' denotes the linear local field inside the graded inclusions or the host.

In view of the existence of nonlinearity inside the anisotropic graded particles, χ_e is given by [66, 136]

$$\chi_e E_0^4 = \frac{1}{V} \sum_{ijkl} \int_V \chi_{ijkl} E_{\text{lin},i} E_{\text{lin},j} E_{\text{lin},k} E_{\text{lin},l}^* dV = f \sum_{ijkl} \langle \chi_{ijkl} E_{\text{lin},i} E_{\text{lin},j} E_{\text{lin},k} E_{\text{lin},l} \rangle. \tag{5.106}$$

Here $E_{\text{lin},i}$ denotes the Cartesian component of the linear local electric field. Then, just as in isotropic composites [66], both ε_e and χ_e in nonlinear composite media with local dielectric anisotropy can be expressed (to the lowest order in the nonlinearity) in terms of the electric field in the related linear medium as well.

Next, we will develop a nonlinear anisotropic differential effective dipole approximation (NADEDA), so as to derive the equivalent linear dielectric constant $\bar{\varepsilon}(a)$ and third-order nonlinear susceptibility $\bar{\chi}(a)$ of the nonlinear graded inclusions. In this connection, ε_e and χ_e of this anisotropic graded composite media can further be derived in the dilute limit.

B. Nonlinear anisotropic differential effective dipole approximation

To put forth an NADEDA (nonlinear anisotropic differential effective dipole approximation) for graded particles with dielectric anisotropy, we regard the gradation profiles as a multi-shell construction. In detail, we build up the dielectric profile by adding shells gradually [64]. Let us start with an infinitesimal spherical core with linear dielectric constants $\varepsilon_{1r}(r=0) = \varepsilon_{1t}(r=0) = \varepsilon(0)$ and nonlinear susceptibility χ_{ijkl}, and keep on adding shells with the tangential and radial dielectric constant $\varepsilon_{1t}(r)$ and $\varepsilon_{1r}(r)$, and the Cartesian fourth-rank tensorial nonlinear susceptibility χ_{ijkl} (to show the optical nonlinearity enhancement, we always assume χ_{ijkl} to be independent of r), at radius r, until $r = a$ is reached.

At radius r, we have an inhomogeneous spherical particle with spatially varying dielectric constant, which are characterized by the gradation profiles $\varepsilon_{1r}(r)$ and $\varepsilon_{1t}(r)$, and with tensorial nonlinear susceptibility χ_{ijkl}. Then, we can regard such an inhomogeneous particle as an effective *homogeneous* one with the equivalent isotropic dielectric properties $\bar{\varepsilon}(r)$ and $\bar{\chi}(r)$. Here the *homogeneous* sphere should induce the same dipole moment as the original inhomogeneous sphere. Then, we add to the *homogeneous* particle a spherical shell of infinitesimal thickness dr, with linear dielectric constants, $\varepsilon_{1r}(r)$ and $\varepsilon_{1t}(r)$, and nonlinear susceptibility, χ_{ijkl}. In this situation, the coated inclusions are composed of a spherical core with radius r, linear dielectric constant $\bar{\varepsilon}(r)$ as well as nonlinear susceptibility $\bar{\chi}(r)$, and a shell with outermost radius $r + dr$, linear dielectric constants $\varepsilon_{1r}(r)$ and $\varepsilon_{1t}(r)$, as well as nonlinear susceptibility χ_{ijkl}.

For the graded particles with dielectric anisotropy describe Eq. (5.103), the displacement vector is related to the field, $\mathbf{D} = \bar{\varepsilon}_1(r) \cdot \mathbf{E}$. In view of $\mathbf{E} = -\nabla\Phi$, we have the following electrostatic equation,

$$\nabla \cdot (\bar{\varepsilon}_1(r) \cdot \nabla\Phi) = 0. \tag{5.107}$$

In spherical coordinates, Eq. (5.107) can be cast into,

$$\frac{1}{r^2}\frac{\partial}{\partial r}(r^2\varepsilon_{1r}(r)\frac{\partial\Phi}{\partial r}) + \frac{1}{r^2\sin\theta}\frac{\partial}{\partial\theta}(\sin\theta\varepsilon_{1t}(r)\frac{\partial\Phi}{\partial r}) + \frac{1}{r^2\sin^2\theta}\frac{\partial}{\partial\psi}(\varepsilon_{1t}(r)\frac{\partial\Phi}{\partial\psi}) = 0. \tag{5.108}$$

Let's consider the composite where the coated inclusions are randomly embedded in the linear host medium. Under the quasi-static approximation, we can readily obtain the linear electric potentials inside the core, shell and host medium by solving Eq. (5.108),

$$\phi_c = -E_0 A_0' R \cos\theta, R < r,$$

$$\phi_s = -E_0\left(B_0'(r+dr)^{1-\delta}R^\delta - \frac{C_0'r^{(1+2\delta)}(r+dr)^{1-\delta}}{R^{\delta+1}}\right)\cos\theta, r < R < r+dr,$$

$$\phi_h = -E_0\left(R - \frac{D_0'(r+dr)^3}{R^2}\right)\cos\theta, R > r+dr. \tag{5.109}$$

Here the four unknown parameters A_0', B_0', C_0' and D_0' can be determined by applying the appropriate boundary conditions on the interfaces. As a result, we obtain

$$A_0' = \frac{3(1+2\delta)\varepsilon_2\varepsilon_{1r}(r)\lambda^{(\delta-1)/3}}{Q}, B_0' = \frac{3\varepsilon_2[\bar{\varepsilon}(r)+(1+\delta)\varepsilon_{1r}(r)]}{Q},$$

$$C_0' = \frac{3\varepsilon_2[\bar{\varepsilon}(r)-\delta\cdot\varepsilon_{1r}(r)]}{Q},$$

$$D_0' = \frac{[\delta\cdot\varepsilon_{1r}(r)-\varepsilon_2][\bar{\varepsilon}(r)+(1+\delta)\varepsilon_{1r}(r)]+\lambda^{(1+2\delta)/3}[\varepsilon_2+(1+\delta)\varepsilon_{1r}(r)]\vartheta}{Q},$$

with $\vartheta = [\bar{\varepsilon}(r) - \delta\cdot\varepsilon_{1r}(r)]$, $\lambda \equiv [r/(r+dr)]^3$, $\delta \equiv -1/2 + \sqrt{1/4 + 2\varepsilon_{1t}(r)/\varepsilon_{1r}(r)}$, and

$$Q = [\delta\cdot\varepsilon_{1r}(r)+2\varepsilon_2][\bar{\varepsilon}(r)+(1+\delta)\varepsilon_{1r}(r)]+\lambda^{(1+2\delta)/3}[(1+\delta)\varepsilon_{1r}(r)-2\varepsilon_2][\bar{\varepsilon}(r)-\delta\cdot\varepsilon_{1r}(r)].$$

If $\varepsilon_{1t}(r) = \varepsilon_{1r}(r)$, the physical parameter $\delta = 1$, and then Eq. (5.109) degenerates to the isotropic form.

The effective (overall) linear dielectric constant of the system is determined by the dilute-limit expression [71]

$$\varepsilon_e = \varepsilon_2 + 3p\varepsilon_2 D_0', \qquad (5.110)$$

where p is the volume fraction of the graded particles with radius r. The equivalent dielectric constant $\bar{\varepsilon}(r+dr)$ for the graded particles with radius $r+dr$ can be self-consistently obtained by the vanishing of the dipole factor D by replacing ε_2 with $\bar{\varepsilon}(r+dr)$. Taking the limit $dr \to 0$ and keeping to the first order in dr, we obtain

$$
\begin{aligned}
\bar{\varepsilon}(r+dr) &= \varepsilon_{1r}(r)\left\{1+\frac{\bar{\varepsilon}(r)[(\delta-1)+(\delta+2)\vartheta']+\varepsilon_{1r}(r)[(\delta^2-1)-(\delta+2)\delta\vartheta']}{\bar{\varepsilon}(r)(1-\vartheta')+\varepsilon_{1r}(r)[\delta+1+\delta\cdot\vartheta']}\right\} \\
&= \bar{\varepsilon}(r)+\frac{[\delta\varepsilon_{1r}(r)-\bar{\varepsilon}(r)][(\delta+1)\varepsilon_{1r}(r)+\bar{\varepsilon}(r)]}{r\varepsilon_{1r}(r)}dr, \qquad (5.111)
\end{aligned}
$$

with $\vartheta' = \lambda^{(1+2\delta)/3}$ Thus, we have the differential equation for the equivalent dielectric constant $\bar{\varepsilon}(r)$,

$$\frac{d\bar{\varepsilon}(r)}{dr} = \frac{\delta(\delta+1)[\varepsilon_{1r}(r)]^2 - \bar{\varepsilon}(r)\varepsilon_{1r}(r)-[\bar{\varepsilon}(r)]^2}{r\varepsilon_{1r}(r)}. \qquad (5.112)$$

Note that Eq. (5.112) is just the Tartar formula, which was derived for assemblages of spheres with varying radial and tangential conductivity [43]. If ε_{1r} is independent of r, namely $\varepsilon_{1r} = \varepsilon_1$, we have $\delta = 1$ due to isotropic property at $r - 0$, and then Eq. (5.112) predicts $\bar{\varepsilon}(r) = \varepsilon_1$, as expected.

Next, we speculate on how to derive the equivalent nonlinear susceptibility $\bar{\chi}(r)$. After applying Eq. (5.106) to the coated particles with radius $r+dr$, we have

$$
\begin{aligned}
\bar{\chi}(r+dr)\frac{\langle|E|^2 E^2\rangle_{R\leq r+dr}}{|E_0|^2 E_0^2} &= \lambda\bar{\chi}(r)\frac{\langle|E|^2 E^2\rangle_{R\leq r}}{|E_0|^2 E_0^2}+(1-\lambda) \\
&\quad \times \frac{\sum_{ijkl}\langle\chi_{ijkl}E_iE_jE_kE_l^*\rangle_{r<R\leq r+dr}}{|E_0|^2 E_0^2}. \qquad (5.113)
\end{aligned}
$$

As $dr \to 0$, the left-hand side of the above equation admits

$$
\begin{aligned}
\bar{\chi}(r+dr)\frac{\langle|E|^2 E^2\rangle_{R\leq r+dr}}{|E_0|^2 E_0^2} &= \bar{\chi}(r+dr)\left|\frac{3\varepsilon_2}{\bar{\varepsilon}(r+dr)+2\varepsilon_2}\right|^2\left(\frac{3\varepsilon_2}{\bar{\varepsilon}(r+dr)+2\varepsilon_2}\right)^2 \\
&= \bar{\chi}(r)|K|^2 K^2 - dr\bar{\chi}(r)|K|^2 K^2\left[\frac{3d\bar{\varepsilon}(r)/dr}{2\varepsilon_2+\bar{\varepsilon}(r)}\right. \\
&\quad \left.+\left(\frac{d\bar{\varepsilon}(r)/dr}{2\varepsilon_2+\bar{\varepsilon}(r)}\right)^*\right]+|K|^2 K^2\frac{d\bar{\chi}(r)}{dr}\cdot dr, \qquad (5.114)
\end{aligned}
$$

with $K = (3\varepsilon_2)/[\bar{\varepsilon}(r)+2\varepsilon_2]$. The first part of the right-hand side of Eq. (5.113) is written as

$$\lambda\frac{\bar{\chi}(r)\langle|E|^2 E^2\rangle_{R\leq r}}{|E_0|^2 E_0^2} = \bar{\chi}(r)|K|^2 K^2\left[1+(3y+y^*-3)\frac{dr}{r}\right], \qquad (5.115)$$

where

$$y = \frac{[(1+\delta)\varepsilon_{1r}(r)-2\varepsilon_2][\bar{\varepsilon}(r)-\delta\cdot\varepsilon_{1r}(r)]}{\varepsilon_{1r}(r)[\bar{\varepsilon}(r)+2\varepsilon_2]}-\delta+1.$$

The term $U \equiv (\sum_{ijkl}\langle\chi_{ijkl}(r)E_iE_jE_kE_l^*\rangle_{r<R\leq r+dr})/(|E_0|^2E_0^2)$ in Eq. (5.113) is written as,

$$
\begin{aligned}
U &= [(\chi_{xxyy}+\chi_{yxxy}+\chi_{xyxy}+\chi_{xyyx}+\chi_{yxyx}+\chi_{yyxx}+3\chi_{xxxx}+3\chi_{yyyy})U_{p1} \\
&\quad (\chi_{xxzz}+\chi_{xzxz}+\chi_{zxxz}+\chi_{zyyz}+\chi_{yzyz}+\chi_{yyzz})U_{p2}+(\chi_{zzxx}+\chi_{xzzx} \\
&\quad +\chi_{zxzx}+\chi_{yzzy}+\chi_{zyzy}+\chi_{zzyy})U_{p3}+\chi_{zzzz}U_{p4}]\frac{|K|^2K^2}{315},
\end{aligned} \tag{5.116}
$$

where

$$
\begin{aligned}
U_{p1} &= [B_2(\delta-1)+C_2(2+\delta)]^3\cdot[B_2^*(\delta-1)+C_2^*(2+\delta)], \\
U_{p2} &= [B_2(\delta-1)+C_2(2+\delta)]^2\cdot[|C_2|^2(5+2\delta+5\delta^2) \\
&\quad +(C_2B_2^*+B_2C_2^*)(-4+5\delta+5\delta^2)+|B_2|^2(8+8\delta+5\delta^2)], \\
U_{p3} &= [B_2^*(\delta-1)+C_2^*(2+\delta)]\cdot[B_2^3(-8+3\delta^2+5\delta^3)+ \\
&\quad 3B_2^2C_2(8+2\delta+6\delta^2+5\delta^3)+3B_2C_2^2(-7+5\delta+9\delta^2+5\delta^3) \\
&\quad +C_2^3(10+9\delta+12\delta^2+5\delta^3)], \\
U_{p4} &= [B_2^2|B_2|^2(128+64\delta+48\delta^2+40\delta^3+35\delta^4) \\
&\quad +B_2^2(3C_2B_2^*+B_2C_2^*)(-112-8\delta+30\delta^2+55\delta^3+35\delta^4) \\
&\quad +3B_2C_2(C_2B_2^*+B_2C_2^*)(104+4\delta+39\delta^2+70\delta^3+35\delta^4) \\
&\quad +C_2^2(C_2B_2^*+3B_2C_2^*)(-94+43\delta+75\delta^2+85\delta^3+35\delta^4) \\
&\quad +C_2^2|C_2|^2(107+52\delta+138\delta^2+100\delta^3+35\delta^4)],
\end{aligned}
$$

with

$$
B_2 = \frac{\bar{\varepsilon}(r)+(1+\delta)\varepsilon_{1r}(r)}{(1+2\delta)\varepsilon_{1r}(r)} \quad\text{and}\quad C_2 = \frac{\bar{\varepsilon}(r)-\delta\cdot\varepsilon_{1r}(r)}{(1+2\delta)\varepsilon_{1r}(r)}.
$$

Substituting Eqs. (5.114), (5.115) and (5.116) into Eq. (5.113), we have a differential equation for the equivalent nonlinear susceptibility $\bar{\chi}(r)$, namely,

$$
\begin{aligned}
\frac{d\bar{\chi}(r)}{dr} &= \bar{\chi}(r)\left[\frac{3d\bar{\varepsilon}(r)/dr}{2\varepsilon_2+\bar{\varepsilon}(r)}+\left(\frac{d\bar{\varepsilon}(r)/dr}{2\varepsilon_2+\bar{\varepsilon}(r)}\right)^*\right] \\
&\quad +\bar{\chi}(r)\cdot\frac{3y+y^*-3}{r}+\frac{3}{r}\cdot\frac{U}{|K|^2K^2}.
\end{aligned} \tag{5.117}
$$

From Eqs. (5.116) and (5.117), it is evident that χ_{ijkl} does not contribute to the equivalent nonlinear susceptibility, except for the cases with equal pair indices.

So far, the equivalent $\bar{\varepsilon}(r)$ and $\bar{\chi}(r)$ of the anisotropic graded spherical particles with radius r can be calculated, at least numerically, by solving the differential equations Eqs. (5.112) and (5.117), as long as $\varepsilon_{1r}(r)$, $\varepsilon_{1t}(r)$ (dielectric-constant gradation profiles) and χ_{ijkl} are given. Here we would like to mention that, even though χ_{ijkl} is independent of r, the equivalent $\bar{\chi}(r)$ should still be dependent on r. This is because both $\varepsilon_{1r}(r)$ and $\varepsilon_{1t}(r)$ are r-dependent. Moreover, if $\chi_{ijkl}=0$, Eq. (5.113) admits $\bar{\chi}(r)=0$, as expected as well.

To obtain $\bar{\varepsilon}(r=a)$ and $\bar{\chi}(r=a)$, we integrate Eqs. (5.112) and (5.117) numerically, for given initial conditions $\bar{\varepsilon}(r\to0)$ and $\bar{\chi}(r\to0)$. Once $\bar{\varepsilon}(r=a)$ and $\bar{\chi}(r=a)$ are calculated, we can take one step forward to work out the effective linear and nonlinear responses of the whole composite ε_e and χ_e [134],

$$
\varepsilon_e = \varepsilon_2+3\varepsilon_2 f\frac{\bar{\varepsilon}(r=a)-\varepsilon_2}{\bar{\varepsilon}(r=a)+2\varepsilon_2}, \tag{5.118}
$$

and

$$\chi_e = f\bar{\chi}(r=a) \left| \frac{3\varepsilon_2}{\bar{\varepsilon}(r=a)+2\varepsilon_2} \right|^2 \left(\frac{3\varepsilon_2}{\bar{\varepsilon}(r=a)+2\varepsilon_2} \right)^2. \tag{5.119}$$

C. Exact solution for linear gradation profiles

Based on the first-principles approach, the potentials within the graded inclusions and the host medium can be exactly obtained, when the dielectric gradation profiles are the linear radial functions with small slopes, i.e., $\varepsilon_{1r}(r) = \varepsilon(0) + g(r/a)$ and $\varepsilon_{1t}(r) = \varepsilon(0) + h(r/a)$. Here $g \,[< a\varepsilon(0)]$ and h are two constants, and $\varepsilon(0)$ denotes the linear dielectric constant at radius $r=0$.

The potentials within the graded spheres and the host medium are respectively given by

$$\phi_c(r,\theta) = -E_0 A_1 \sum_{k=0}^{\infty} C_k \left(\frac{gr}{a\varepsilon(0)} \right)^{k+1} \cos\theta, \quad r < a,$$

$$\phi_m(r,\theta) = -E_0 r \cos\theta + \frac{D_1}{r^2} E_0 \cos\theta, \quad r > a, \tag{5.120}$$

where the coefficients A_1 and D_1 have the following forms

$$A_1 = \frac{3\varepsilon_2 a}{(\varepsilon(0)+g)v_2 + 2\varepsilon_2 v_1} \quad \text{and} \quad D_1 = \frac{(\varepsilon(0)+g)v_2 - \varepsilon_2 v_1}{(\varepsilon(0)+g)v_2 + 2\varepsilon_2 v_1} a^3.$$

Here v_1 and v_2 are given by

$$v_1 = \sum_{k=0}^{\infty} C_k \left(\frac{g}{\varepsilon(0)} \right)^{k+1} \quad \text{and} \quad v_2 = \sum_{k=0}^{\infty} C_k(k+1) \left(\frac{g}{\varepsilon(0)} \right)^{k+1},$$

with C_k satisfying the following recurrent relation,

$$C_{k+1} = -\frac{(k+1)(k+3) - 2h/g}{(k+2)(k+3) - 2} C_k.$$

The local electric field inside the anisotropic graded inclusions can be derived from the relation $\mathbf{E} = -\nabla\phi$, and we have

$$\mathbf{E}_c = A_1 E_0 \sum_{k=0}^{\infty} C_k \left(\frac{g}{a\varepsilon(0)} \right)^{k+1} r^k \left[(k+1)\cos\theta\mathbf{e}_r - \sin\theta\mathbf{e}_\theta \right]$$

$$= A_1 E_0 \sum_{k=0}^{\infty} C_k \left(\frac{g}{a\varepsilon(0)} \right)^{k+1} r^k \{ k\cos\theta\sin\theta\cos\phi\mathbf{e}_x + k\cos\theta\sin\theta\sin\phi\mathbf{e}_y$$

$$\left[(k+1)\cos^2\theta + \sin^2\theta \right] \mathbf{e}_z \}. \tag{5.121}$$

Then, the corresponding displacement admits

$$\mathbf{D}_c = \bar{\varepsilon}_1(r)\mathbf{E}_c = A_1 E_0 \sum_{k=0}^{\infty} C_k \left(\frac{g}{a\varepsilon(0)} \right)^{k+1} r^k$$

$$\times [\varepsilon_{1r}(r)(k+1)\cos\theta\mathbf{e}_r - \varepsilon_{1t}(r)\sin\theta\mathbf{e}_\theta]$$

$$= A_1 E_0 \sum_{k=0}^{\infty} C_k \left(\frac{g}{a\varepsilon(0)} \right)^{k+1} r^k \{ [(\varepsilon_{1r}(r)(k+1) - \varepsilon_{1t}(r)] \cos\theta\sin\theta\cos\phi\mathbf{e}_x +$$

$$[\varepsilon_{1r}(k+1) - \varepsilon_{1t}(r)] \cos\theta\sin\theta\sin\phi\mathbf{e}_y$$

$$+ \left[\varepsilon_{1r}(r)(k+1)\cos^2\theta + \varepsilon_{1t}(r)\sin^2\theta \right] \mathbf{e}_z \}. \tag{5.122}$$

where $\mathbf{e}_r$ and $\mathbf{e}_\theta$ ($\mathbf{e}_x$, $\mathbf{e}_y$ and $\mathbf{e}_z$) are the unix vectors in spherical coordinates (Cartesian coordinates).

In the dilute limit, from Eq. (5.105), we can obtain the effective linear dielectric constant as

$$
\begin{aligned}
\varepsilon_e &= \varepsilon_2 + \frac{1}{VE_0}\int_{\Omega_i}\left(\bar{\varepsilon}_1(r)\cdot\mathbf{E}-\varepsilon_2\mathbf{E}\right)\cdot\mathbf{e}_z dV \\
&= \varepsilon_2 + 3f\varepsilon_2\frac{[\varepsilon(0)-\varepsilon_2]v_1+gv_3+2hv_4}{[\varepsilon(0)+g]v_2+2\varepsilon_2 v_1},
\end{aligned}
\tag{5.123}
$$

where

$$
v_3 = \sum_{k=0}^{\infty}C_k\frac{1+k}{4+k}\left(\frac{g}{\varepsilon(0)}\right)^{k+1} \quad \text{and} \quad v_4 = \sum_{k=0}^{\infty}C_k\frac{1}{4+k}\left(\frac{g}{\varepsilon(0)}\right)^{k+1}.
$$

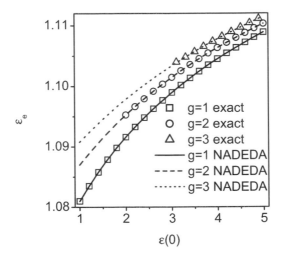

Figure 5.31. The effective linear dielectric constant (ε_e) versus the dielectric constant of the spherical core with radius $r = 0^+$ [$\varepsilon(0)$], for the linear dielectric gradation profiles with various radial gradients g. Parameters: volume fraction $f = 0.05$, the tangential gradient $h = 8$. Lines: numerical results obtained from the NADEDA [Eqs. (5.112) and (5.118)]; Symbols: exact results predicted by the first-principles approach [Eq. (5.123)]. Note that the exact results are available for $\varepsilon(0) > g$. After Ref. [137].

On the other hand, the substitution of Eq. (5.121) into Eq. (5.106) yields

$$
\begin{aligned}
\chi_e &= \frac{1}{V}\sum_{ijkl}\int_{\Omega_i}\chi_{ijkl}E_iE_jE_kE_l^* dV \\
&= f[(\chi_{xxyy}+\chi_{yxxy}+\chi_{xyxy}+\chi_{xyyx}+\chi_{yxyx}+\chi_{yyxx}+3\chi_{xxxx}+3\chi_{yyyy})U_{q1} \\
&\quad (\chi_{xxzz}+\chi_{xzxz}+\chi_{zxxz}+\chi_{yyzz}+\chi_{yzyz}+\chi_{zyyz})U_{q2}+(\chi_{zzxx}+\chi_{zxzx} \\
&\quad +\chi_{xzzx}+\chi_{zzyy}+\chi_{yzzy}+\chi_{zyzy})U_{q3}+\chi_{zzzz}U_{q4}],
\end{aligned}
\tag{5.124}
$$

where

$$U_{q1} = \frac{1}{105}|A_2|^2 A_2^2 \sum_{k_1=0}^{\infty}\sum_{k_2=0}^{\infty}\sum_{k_3=0}^{\infty}\sum_{k_4=0}^{\infty}\left\{\left[C_{k_1}C_{k_2}C_{k_3}C_{k_4}\left(\frac{g}{\varepsilon(0)}\right)^{k_1+k_2+k_3+3}\right]\right.$$

$$\left.\cdot\left[\left(\frac{g}{\varepsilon(0)}\right)^{k_4+1}\right]^* \frac{k_1 k_2 k_3 k_4}{3+k_1+k_2+k_3+k_4}\right\},$$

$$U_{q2} = 3|A_2|^2 A_2^2 \sum_{k_1=0}^{\infty}\sum_{k_2=0}^{\infty}\sum_{k_3=0}^{\infty}\sum_{k_4=0}^{\infty}\left\{\left[C_{k_1}C_{k_2}C_{k_3}C_{k_4}\left(\frac{g}{\varepsilon(0)}\right)^{k_1+k_2+k_3+3}\right]\right.$$

$$\left.\cdot\left[\left(\frac{g}{\varepsilon(0)}\right)^{k_4+1}\right]^* \frac{k_1 k_2}{3+k_1+k_2+k_3+k_4}\left[\frac{1}{15}+\frac{1}{35}(k_3+k_4)+\frac{1}{63}k_3 k_4\right]\right\},$$

$$U_{q3} = 3|A_2|^2 A_2^2 \sum_{k_1=0}^{\infty}\sum_{k_2=0}^{\infty}\sum_{k_3=0}^{\infty}\sum_{k_4=0}^{\infty}\left\{\left[C_{k_1}C_{k_2}C_{k_3}C_{k_4}\left(\frac{g}{\varepsilon(0)}\right)^{k_1+k_2+k_3+3}\right]\right.$$

$$\left.\cdot\left[\left(\frac{g}{\varepsilon(0)}\right)^{k_4+1}\right]^* \frac{k_3 k_4}{3+k_1+k_2+k_3+k_4}\left[\frac{1}{15}+\frac{1}{35}(k_1+k_2)+\frac{1}{63}k_1 k_2\right]\right\},$$

$$U_{q4} = 3|A_2|^2 A_2^2 \sum_{k_1=0}^{\infty}\sum_{k_2=0}^{\infty}\sum_{k_3=0}^{\infty}\sum_{k_4=0}^{\infty}\left\{\left[C_{k_1}C_{k_2}C_{k_3}C_{k_4}\left(\frac{g}{\varepsilon(0)}\right)^{k_1+k_2+k_3+3}\right]\right.$$

$$\cdot\left[\left(\frac{g}{\varepsilon(0)}\right)^{k_4+1}\right]^* \frac{1}{3+k_1+k_2+k_3+k_4}\left[1+\frac{1}{3}\sum_{i=1}^{4}k_i+\frac{1}{5}\sum_{i=1}^{3}\sum_{j=i+1}^{4}k_i k_j\right.$$

$$\left.\left.\frac{1}{7}\sum_{i=1}^{2}\sum_{j=i+1}^{3}\sum_{l=j+1}^{4}k_i k_j k_l+\frac{1}{9}k_1 k_2 k_3 k_4\right]\right\},$$

with $A_2 = (3\varepsilon_2)/\{[\varepsilon(0)+g]v_2+2\varepsilon_2 v_1\}$.

To illustrate the NADEDA, we first perform numerical calculations for the linear dielectric gradation profiles, that is, $\varepsilon_{1r}(r) = \varepsilon(0)+gr/a$ (radial dielectric constant), and $\varepsilon_{1t}(r) = \varepsilon(0)+hr/a$ (tangential dielectric constant). In this situation, the exact results for ε_e and χ_e exist, and thus it allows us to show the correctness of the NADEDA. For model calculations, we set $h > g$ (Note that our formulae can still be used for $h \leq g$). For the NADEDA, we numerically integrate Eqs. (5.112) and (5.117) by using *Mathematica* with the initial radius $r = 0.001$.

In Fig. 5.31, the effective linear dielectric constant (ε_e) is plotted as a function of the dielectric constant of anisotropic graded particles at radius $r = 0$ [$\varepsilon(0)$], for various gradients h and g. It is shown that ε_e increases monotonically with the increase of $\varepsilon(0)$. Moreover, increasing the gradient g causes ε_e to increase as well. This can be understood by the fact that the increases of both $\varepsilon(0)$ and g lead to the increase of the equivalent dielectric constant $\bar{\varepsilon}(a)$ of the graded particles, thus increasing the effective response of the whole system. For ε_e, the NADEDA shows good agreement with the first-principles approach.

Next, we investigate the effective third-order nonlinear susceptibility. Let's set the tensorial dielectric susceptibility of the particles to be independent of r, in an attempt to focus on the nonlinearity enhancement. As a result, it is shown that the nonlinearity enhancement decreases with the increase of $\varepsilon(0)$ and g. As mentioned above, for larger ε_e and

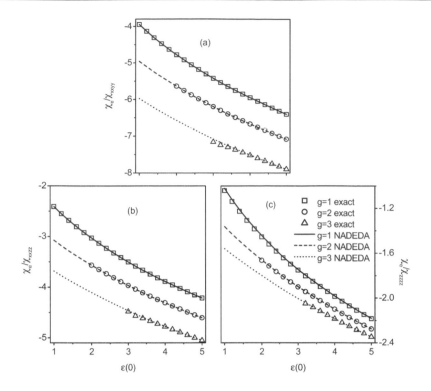

Figure 5.32.   (a) χ_e/χ_{xxyy} versus $\varepsilon(0)$, for the linear dielectric gradation profiles with various g, at $h = 8$. Here χ_{xxyy} is the only nonzero component of the tensorial nonlinear susceptibility.   Lines: numerical results obtained from the NADEDA [Eqs. (5.112), (5.117) and (5.119)]; Symbols: exact results predicted by the first-principles approach [Eq. (5.124)]. (b) Same as (a), but χ_e/χ_{xxzz} versus $\varepsilon(0)$, with χ_{xxzz} being the only nonzero component.   (c) Same as (a), but χ_e/χ_{zzzz} versus $\varepsilon(0)$, with χ_{zzzz} being the only nonzero component. After Ref. [137].

g, the graded inclusions possess a larger equivalent dielectric constant, and hence the i-th Cartesian component of the local field should become more weak accordingly. Then, the weaker effective nonlinear susceptibility is obtained. As displayed in Fig. 5.32, we show three typical cases of nonlinearity enhancement. Here, all the physical parameters in use are real, and thus the nonlinearity enhancement for χ_{zzxx} (the only nonzero component) is the same as that for χ_{xxzz}. Moreover, for other nonzero components of the tensorial nonlinear susceptibility, the nonlinearity enhancement will be the same as one of these shown in Fig. 5.32. For example, Fig. 5.32(a) can also show the nonlinearity enhancement for $3\chi_{xxxx}$, $3\chi_{yyyy}$, χ_{yxxy}, etc. Again, the excellent agreement is numerically demonstrated between the first-principles approach and the NADEDA [Eqs. (5.112), (5.117) and (5.119)].

In what follows, we shall investigate the surface plasma resonance effect on the nonlinear metal-dielectric composite. As a model calculation, we assume the radial and tangential dielectric constants for the graded metal particles to be Drude-like, namely,

$$\varepsilon_{1r}(r) = 1 - \frac{\omega_{pr}^2(r)}{\omega(\omega+i\gamma)} \quad \text{and} \quad \varepsilon_{1t}(r) = 1 - \frac{\omega_{pt}^2(r)}{\omega(\omega+i\gamma)} \tag{5.125}$$

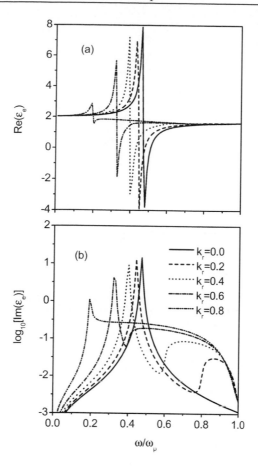

Figure 5.33. The real and imaginary parts of ε_e versus frequency ω/ω_p, for various k_r. Parameters: $\omega_p/\gamma = 0.01$ and $k_t = 0$. After Ref. [137].

where $\omega_{pr}(r)$ and $\omega_{pt}(r)$ are the radius-dependent radial and tangential plasma frequencies, respectively, and γ is the damping coefficient. For the linear dielectric host, we choose $\varepsilon_2 = 1.77$ (a typical dielectric constant of water). We further assume $\omega_{pr}(r)$ and $\omega_{pt}(r)$ to be

$$\omega_{pr}(r) = \omega_p(1 - k_r \cdot \frac{r}{a}), \quad \text{and} \quad \omega_{pt}(r) = \omega_p(1 - k_t \cdot \frac{r}{a}), \quad r < a. \tag{5.126}$$

The above form is quite physical for $0 < k_r(k_t) < 1$, since the center of grains can be better metallic so that $\omega_p(r)$ is larger, while the boundary of the grain may be poorer metallic so that $\omega_p(r)$ is much smaller. In fact, such a variation can also appear owing to the temperature effect [123]. Moreover, we choose $k_t \leq k_r$, in view of the strong metallic behavior in the tangential direction.

Fig. 5.33 displays the real and imaginary parts of effective dielectric constant ε_e as a function of the incident angular frequency ω/ω_p. For $k_r = 0$, there exists a frequency region, where the real part of the effective dielectric constant is negative. With increasing k_r, this region becomes narrow generally, in accompanied with less negative $\text{Re}(\varepsilon_e)$ [see Fig. 5.33(a)]. This is due to the fact that increasing k_r decreases the influence of the metallic

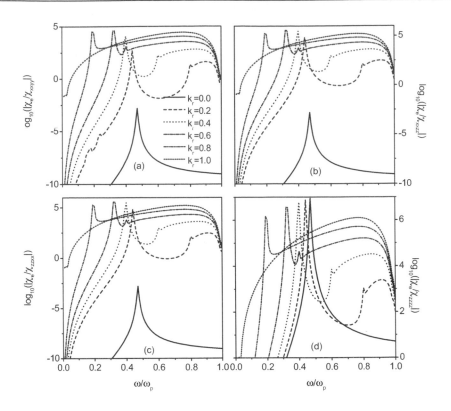

Figure 5.34. Same as Fig. 5.33, but (a) $|\chi_e/\chi_{xxyy}|$ versus ω/ω_p, with χ_{xxyy} being the only nonzero component. (b) $|\chi_e/\chi_{xxzz}|$ versus ω/ω_p, with χ_{xxzz} being the only nonzero component. (c) $|\chi_e/\chi_{zzxx}|$ versus ω/ω_p, with χ_{zzxx} being the only nonzero component. (d) $|\chi_e/\chi_{zzzz}|$ versus ω/ω_p, with χ_{zzzz} being the only nonzero component. After Ref. [137].

behavior [owing to the decrease of $\omega_{pr}(r)$]. In the mean time, the sharp peak for $\text{Im}(\varepsilon_e)$ turn weak with k_r [see Fig. 5.33(b)]. Furthermore, for $k_r \neq 0$, the continuous resonant bands in the high frequency region appear always, and this region becomes more broad as k_r increases. In this case, the appearance of the resonant bands results from the radius-dependent plasma frequency $\omega_p(r)$. This phenomenon has already been observed, when a shell model [74] or nonspherical model [138] was taken into account. In our previous works [74, 138], a broad continuous spectrum is shown to be around the larger pole in the corresponding spectral density function. Here, the graded particles under consideration can be regarded as a construction of multi shells, which hence should be expected to yield the broader spectra for the optical absorption [$\text{Im}(\varepsilon_e)$]. In addition, we note that, as k_r increases, both the surface plasma frequency and the center of resonant bands are red-shifted. In particular, for larger k_r, the resonant bands can become broader, owing to strong inhomogeneity inside the particles.

Then, we speculate on how gradation and anisotropy affect the optical nonlinearity enhancement in metal-dielectric composites. As shown in Fig. 5.34, no matter which component of the nonlinear susceptibility tensor is nonzero, χ_e can be substantially enhanced within a certain frequency region. In particular, this enhancement becomes quite strong for

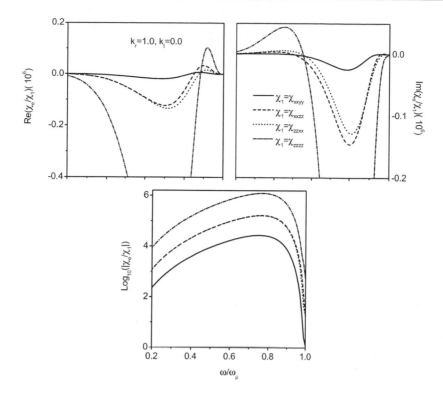

Figure 5.35. The real and imaginary parts, and the modulus of the optical nonlinearity enhancement for $k_r = 1$ and $k_t = 0$. After Ref. [137].

χ_{zzzz} (the only nonzero component). In fact, the physical origin of this huge enhancement is the large increase in the local field component F_z. In addition, the nonlinearity enhancement will become more strong, for the system with a larger k_r which is related to a higher contrast between ε_{1t} and ε_{1r}. For example, $|\chi_e/\chi_{zzzz}| > 10^4$ in the frequency region $0.2 \le \omega/\omega_p \le 1.0$ for $k_r = 1$. From Fig. 5.34, we also find that the optical nonlinearity enhancement obtained for four nonzero components, respectively, displays the similar qualitative behaviors. This should be in contrast to those observed in a a polycrystalline quasi-one-dimensional conductor [134, 136, 139], where the effective optical nonlinearity for four elements of the nonlinear susceptibility tensor exhibit quite different behaviors [134] (the differences become more distinct by using spectral representation approximation [136]). Actually, the differences result from two different kinds of dielectric anisotropy (and hence two different kinds of tensorial dielectric constants) under consideration. Here we focus on the particles with spatially varying, but spherically symmetric, dielectric anisotropy, whereas, in the Refs. [134, 136, 139], the authors studied uniaxial anisotropy in the Cartesian coordinate system.

Although the optical nonlinearity enhancements for four typical nonzero components of the nonlinear susceptibility (χ_{ijkl}) take on quite similar behaviors, their contributions to the magnitude of the effective optical nonlinearity are different (see Fig. 5.35). As shown in Fig. 5.35, the strongest (weakest) nonlinearity enhancement occurs for the case with χ_{zzzz} (χ_{xxyy}) being only nonzero component. Moreover, the differences between the two cases of

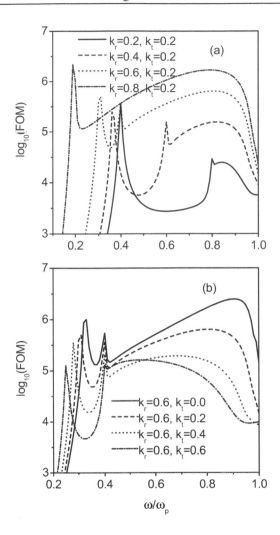

Figure 5.36. The figure of merit $\equiv |\chi_e/\chi_{zzzz}|/\mathrm{Im}(\varepsilon_e)$ versus ω/ω_p, with χ_{zzzz} being the only nonzero component. After Ref. [137].

$\chi_{xxzz} \neq 0$ only and $\chi_{zzxx} \neq 0$ only are clearly shown for $\mathrm{Re}(\chi_e/\chi_1)$ and $\mathrm{Im}(\chi_e/\chi_1)$.

For practical applications, a most useful parameter is the FOM, which is defined as the ratio of $|\chi_e|$ to $\mathrm{Im}(\varepsilon_e)$. In Fig. 5.36, we investigate the figure of merit. Here the only nonzero component is assumed to be χ_{zzzz}. We find that the increase of k_r (namely, the rapid decrease of the radial metallic behavior) results in a large enhancement of the FOM, especially in the high frequency range [see Fig. 5.36(a)]. However, the increase of k_t (i.e., the rapid decrease of the tangential metallic behavior) causes the FOM in the high-frequency region to decrease [see Fig. 5.36(b)]. For instance, we attain FOM$> 10^5$ (which is quite large) in the frequency region $0.3 \leq \omega/\omega_p \leq 1.0$ for $k_r = 0.6$ and $k_t = 0$. Therefore, it is possible to achieve a large figure of merit by introducing the radial gradation and keeping the tangential dielectric constant unchanged.

We have developed an NADEDA (nonlinear anisotropic differential effective dipole approximation) to investigate the effective linear dielectric constant and third-order nonlinear susceptibility of composite media consisting of nonlinear inclusions with spatially varying dielectric anisotropy. Alternatively, based on the first-principles approach, we have derived the exact expressions for ε_e and χ_e, for the linear dielectric-constant profiles with small slopes. To our interest, excellent agreement is found between the approximation results (NADEDA) and the exact results (first-principles approach). It is worth noting that exact solutions are very few in composite research, and thus our NADEDA provides an effective way to estimate the effective nonlinear properties in composite media consisting of anisotropic graded inclusions.

An an application, we apply the NADEDA to study the surface plasma resonance effect on the effective linear dielectric constant, the optical nonlinearity enhancement and the figure of merit in metal-dielectric composites, in which the metal particles possess the tensorial dielectric constants with dielectric gradation profiles. It is found that the gradation profiles in radial dielectric constants are a useful way to control the local-field effects, thus being able to enhance the figure of merit hugely.

The present methods are strictly valid in the dilute limit. The presence of both gradation and dielectric anisotropy is shown to be helpful to achieve the large figure of merit, but unable to realize the separation of the absorption peak from the nonlinearity enhancement peak. In this regard, we may intentionally manipulate composite microstructures, e.g., by using the shape distribution of graded inclusions [125], and by using fractal [73] and anisotropic microstructures [102] with large volume fractions. When the volume fraction of graded inclusions is large, percolation behaviors can occur. To this end, the further broadening of the enhancement peak as well as the desired separation of the optical absorption from the nonlinearity peak is expected to be realized.

Our consideration can be generalized to the nonlinear composites of anisotropic graded inclusions, which is subject to an external alternating current (AC) electric field. For a sinusoidal applied field, the electric response in the composites will generally consist of AC fields at frequencies of high-order harmonics. Initial results show that the fundamental and third-order harmonic AC responses are sensitive to the dielectric gradation profiles as well as anisotropy. Thus, by measuring the AC responses of the anisotropic graded composites, it is possible to perform a real-time monitoring of the fabrication process of the gradation profiles within the particles.

To sum up, we have put forth an NADEDA (nonlinear anisotropic differential effective dipole approximation), in an attempt to discuss the effects of gradation as well as anisotropy on the optical properties of composite media. For the linear dielectric-constant profiles, the NADEDA has been numerically demonstrated in good agreement with the first-principles approach. To our great interest, both the huge nonlinearity enhancement and the large figure of merit are shown to be achievable by the presence of gradation as well as local anisotropy inside the inclusions.

5. Spectral Representation of Graded Composites

Here we shall present a spectral representation of graded composites. This was motivated by a recent study of the optical absorption spectrum of a graded metallic film [14]. In that work, a broad surface plasmon absorption band was observed in addition to a strong Drude absorption peak at zero frequency. Such a broad absorption band has been shown to be responsible for the enhanced nonlinear optical response as well as an attractive FOM. Yuen *et al.* [102] pointed out that such an absorption spectrum, being related to the imaginary part of the effective dielectric constant, should equally well be reflected in the Bergman-Milton spectral representation of the effective dielectric constant [70, 140, 141].

Bergman-Milton spectral representation was originally developed for calculating the effective dielectric constant and other response functions of two-component composites [70, 140, 141]. However, the two concerned components are all homogeneous. Therefore, it is worth extending the spectral representation to graded composite materials. The work on graded films is just a simple example of a more general graded composite in three dimensions. Spectral representation can be used to help identify the physical origin of the broad absorption band. It turns out that, unlike in the case of homogeneous materials, the characteristic function of a graded composite is a continuous function because of the continuous variation of the dielectric function within the constituent component.

Moreover, we apply our theory to a special case of graded composites, i. e., multilayer material, which is more convenient to fabricate in practice than graded material [142], and many algorithms are now available for designing of multilayer coatings [143, 144]. Thus, the present theory is necessary in the sense that we shall discuss the multilayer effect as the number of layers inside the material increases. In this regard, this should be expected to have practical relevance.

We consider a two-component composite in which graded inclusions of dielectric constant $\varepsilon_1(\mathbf{r})$ are embedded in a homogeneous host medium of dielectric constant ε_2. It is noted that the dielectric constant $\varepsilon_1(\mathbf{r})$ is a gradation profile as a function of the position $\mathbf{r}$. And we will restrict our discussion and calculation to the quasi-static approximation, i. e., $dc/\omega \leq 1$, where d is the characteristic size of the inclusion, c is the speed of light in vacuum and ω is the frequency of the applied field. In the quasi-static approximation, the whole graded structure can be regarded as an effective homogeneous one with effective (overall) linear dielectric constant defined as [66]

$$\varepsilon_e = \frac{1}{V} \int \frac{\mathbf{E} \cdot \mathbf{D}}{E_0^2} dV, \qquad (5.127)$$

where E_0 is the applied electric field along z direction, $\mathbf{E}$ and $\mathbf{D}$ are the local electric field and local displacement, respectively.

The object of the present section is to solve the Laplace's equation

$$\nabla \cdot (\varepsilon(\mathbf{r}) \nabla \phi(\mathbf{r})) = 0 \qquad (5.128)$$

subject to the boundary condition $\phi_0 = -E_0 z$. The dielectric function $\varepsilon(\mathbf{r})$ varies from component to component but has a fixed mathematical expression for a given component. It can be expressed as [70]

$$\varepsilon(\mathbf{r}) = \varepsilon_2 \left[1 - \frac{1}{s} \eta(\mathbf{r}) \right], \qquad (5.129)$$

where $s = [1 - \varepsilon_{\text{ref}}/\varepsilon_2]^{-1}$ is the material parameter and ε_{ref} is some reference dielectric constant in the graded component. The characteristic function $\eta(\mathbf{r})$ is may be written in terms of a real function $f(\mathbf{r})$ as

$$\eta(\mathbf{r}) = \begin{cases} 1 + f(\mathbf{r}) & \text{in inclusion,} \\ 0 & \text{in host,} \end{cases}$$

which accords for the microstructure of graded composites. The function $f(\mathbf{r})$ depends on the specific variation of the dielectric constant in the inclusion component. For homogeneous constituent component, i.e., $f(\mathbf{r}) = 0$, $\eta(\mathbf{r}) = 1$ in the inclusion component, while $\eta(\mathbf{r}) = 0$ in the host medium. For graded systems, $\eta(\mathbf{r})$ can be a continuous function in the inclusion component because of the continuous variation of the dielectric function within the inclusion component. Thus, Eq. (5.127) can be solved

$$\phi(\mathbf{r}) = -E_0 z + \frac{1}{s} \int dV' \eta(\mathbf{r}') \nabla' G(\mathbf{r} - \mathbf{r}') \cdot \nabla' \phi(\mathbf{r}'), \tag{5.130}$$

where $G(\mathbf{r} - \mathbf{r}')$ is a Green's function satisfying:

$$\begin{cases} \nabla^2 G(\mathbf{r} - \mathbf{r}') = -\delta^3(\mathbf{r} - \mathbf{r}') & \text{for } \mathbf{r} \text{ in V,} \\ G = 0 & \text{for } \mathbf{r} \text{ on the boundary.} \end{cases}$$

In order to obtain a solution for Eq. (5.128), we introduce an integral-differential Hermitian operator $\hat{\Gamma}$, which satisfies

$$\hat{\Gamma} \equiv \int dV' \eta(\mathbf{r}') \nabla' G(\mathbf{r} - \mathbf{r}') \cdot \nabla',$$

and define an inner product as

$$\langle \phi | \psi \rangle = \int dV \eta(\mathbf{r}) \nabla \phi^* \cdot \nabla \psi. \tag{5.131}$$

With the above definitions, Eq. (5.130) can be simplified to

$$\phi(\mathbf{r}) = -E_0 z + \frac{1}{s} \hat{\Gamma} \phi(\mathbf{r}).$$

Let s_n and $|\phi_n\rangle$ be the nth eigenvalue and eigenfunction of operator $\hat{\Gamma}$. Then, the generalized eigenvalue problem becomes

$$\nabla \cdot (\eta(\mathbf{r}) \nabla \phi_n) = s_n \nabla^2 \phi_n.$$

The potential $|\phi\rangle$ can be expanded in series of eigenfunctions,

$$|\phi\rangle \equiv \sum_n \left(\frac{s}{s_n - s} \right) \frac{|\phi_n\rangle \langle \phi_n | z \rangle}{\langle \phi_n | \phi_n \rangle}, \tag{5.132}$$

where we choose $E_0 = 1$ for convenience. Since $\eta(\mathbf{r})$ is a real function, the eigenvalues s_n will be real. Also, for graded component, $\eta(\mathbf{r})$ is a continuous function, which will cover the full region, i.e., $-\infty \leq \eta(\mathbf{r}) \leq \infty$. Therefore, the eigenvalues s_n, which depend on the continuously graded microstructure $\eta(\mathbf{r})$, do not lie within the interval $[0, 1]$ but extend to

$-\infty \le s_n \le \infty$ as first pointed by Gu and Gong [145] for three-component composites case. However, eigenvalues s_n still lie in this interval $[0,1]$ for $0 \le \eta(\mathbf{r}) \le 1$.

We are now in the position to find an analytical representation for the effective dielectric constant ε_e according to Eq. (5.127). We take advantage of Green's theorem, the boundary condition $\phi_0 = -z$, and the Maxwell equation $\nabla \cdot \mathbf{D} = 0$ to obtain the effective dielectric constant

$$
\begin{aligned}
\frac{\varepsilon_e}{\varepsilon_2} &= \frac{1}{\varepsilon_2 V} \int (-\nabla\phi) \cdot \mathbf{D} dV \\
&= \frac{-1}{V} \int \hat{\mathbf{z}} \cdot \left[\left(1 - \frac{1}{s}\eta(\mathbf{r})\right) \nabla\phi \right] dV \\
&= 1 + \frac{1}{sV}\langle z|\phi\rangle.
\end{aligned}
\tag{5.133}
$$

If we now introduce the reduced response [70]

$$
F(s) = 1 - \frac{\varepsilon_e}{\varepsilon_2},
\tag{5.134}
$$

and substitute Eq. (5.132) into Eq. (5.133) we find

$$
F(s) = \frac{1}{V} \sum_n \frac{|\langle z|\phi_n\rangle|^2}{\langle \phi_n|\phi_n\rangle} \left(\frac{1}{s - s_n} \right).
$$

We can now express the effective dielectric constant as

$$
\varepsilon_e = \varepsilon_2 \left(1 - \sum_n \frac{f_n}{s - s_n} \right),
\tag{5.135}
$$

where f_n is given by

$$
f_n = \frac{1}{V} \frac{|\langle z|\phi_n\rangle|^2}{\langle \phi_n|\phi_n\rangle}.
$$

Using the above equations, we obtain the following sum rule

$$
\begin{aligned}
\sum_n f_n &= \frac{1}{V}\langle z|z\rangle \\
&= \frac{1}{V} \int dV \eta(\mathbf{r}) \nabla z \cdot \nabla z \\
&= \frac{1}{V} \int dV \eta(\mathbf{r}).
\end{aligned}
\tag{5.136}
$$

It is worth noting that the sum rule will not equal to the volume fraction of inclusion. This is different from the Bergman-Milton spectral representation for two homogeneous systems, in which the sum rule equals to the volume fraction of the inclusion.

When the operator $\hat{\Gamma}$ has a continuous spectrum, Eq. (F.2) should be replaced with the integral form

$$
\varepsilon_e = \varepsilon_2 \left(1 - \int ds' \frac{m(s')}{s - s'} \right),
\tag{5.137}
$$

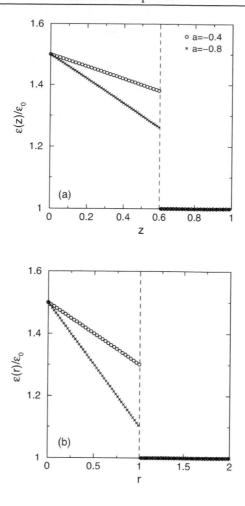

Figure 5.37. (a) Dielectric profile of various graded films at $h = 0.6$. (b) Dielectric proifle of various graded sphere with unit radius. Parameters used: $\varepsilon_2 = 1$ and $s = -2$. After Ref. [57].

where $m(s')$ is the spectral density function. Then, the reduced response becomes

$$F(s) = \int ds' \frac{m(s')}{s - s'}.$$ (5.138)

If we write s as $s + i0^+$, the right side of Eq. (5.138) becomes

$$P \int ds' \frac{m(s')}{s - s'} - i\pi m(s),$$

and thus, $m(s')$ is given through the limiting process

$$m(s') = -\frac{1}{\pi} \text{Im}[F(s' + i0^+)].$$ (5.139)

This final result is identical in form to Bergman's expression for the analogous function in scalar composite materials. However, there are differences in the derivation, namely, the definition of the inner product Eq. (5.131), the continuous graded microstructure $\eta(\mathbf{r})$, the sum rule, as well as the range of eigenvalues s_n.

From Eq. (5.137) it is evident that if the spectral density function $m(s')$ is known, the effective dielectric constant can be obtained accurately, and vice versa. The spectral representation has been used to analyze the effective dielectric properties of composites. Recently, Levy and Bergman [146] also used it in their study of nonlinear optical susceptibility. In this regard, Sheng and coworkers [147] developed a practical algorithm for calculating the effective dielectric constants based on the spectral representation. In what follows, we restrict ourselves to a graded composite both in one dimension and three dimensions, as well as corresponding multilayer composites.

A. Spectral density function of graded composites

1. Spectral density function of a graded composite film

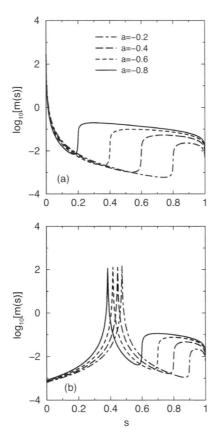

Figure 5.38. (a) Spectral density function of a graded film without an interface, i. e., $h = 1.0$. (b) Spectral density function of a graded film meeting a homogeneous medium at an interface $h = 0.5$, and $\varepsilon_2 = 1$. After Ref. [57].

We consider a graded dielectric film of width L, in which two media meet at a planar interface as shown in Fig. 5.37 (a). The first medium $\varepsilon_1(z)$ varies along $z-$axis, while the second medium ε_2 is homogeneous. We define the graded microstructure as

$$\eta(z) = \begin{cases} 1+az & 0 < z \le h, \\ 0 & h < z < L, \end{cases} \tag{5.140}$$

where a and h are real constants. They can be varied to describe different graded films. Thus, according to Eq. (5.129), the dielectric function of graded film can be expressed as

$$\varepsilon(z) = \varepsilon_2 \left(1 - \frac{\eta(z)}{s} \right). \tag{5.141}$$

Owing to the simple geometry of a graded film, we can use the equivalent capacitance of a series combination to calculate the effective dielectric constant as

$$\frac{1}{\varepsilon_e} = \frac{1}{L} \int_0^L \frac{1}{\varepsilon(z)} dz. \tag{5.142}$$

Substituting Eqs. (5.140) and (5.141) into Eq. (5.142), we obtain

$$\frac{1}{\varepsilon_e} = \frac{1-h}{\varepsilon_2} + \frac{s \left[\ln \left(1 - \frac{\eta(0)}{s} \right) - \ln \left(1 - \frac{\eta(h)}{s} \right) \right]}{a\varepsilon_2},$$

with the assumption $L = 1$.

We are now in a position to extend the Bergman-Milton spectral representation of the effective dielectric constant [70, 140, 141] to a graded film. For a graded system, $\eta(z)$ can be a continuous function in the inclusion medium. Using Eqs. (5.137)–(5.139), we obtain the spectral density function for a graded film as

$$m(s') = -\frac{as' \arg \left(\frac{s-1}{s-ah-1} \right)}{\pi \left[\left(s' \arg \left(\frac{s-1}{s-ah-1} \right) \right)^2 + \left(a(h-1) - s' \ln \left(\frac{s'-1}{s'-ah-1} \right) \right)^2 \right]}, \tag{5.143}$$

where $s' = \text{Re}[s]$ and $\arg[\cdots]$ denote the arguments of complex functions.

2. Spectral density function of a graded sphere

The above theory can be generalized to graded composites in three dimensions. We consider a graded sphere with dielectric constant $\varepsilon_1(r)$ embedded into a homogeneous host medium with dielectric constant ε_2. The dielectric constant of the graded sphere $\varepsilon_1(r)$ varies along the radius r. We can obtain the effective dielectric constant of a graded sphere using the spectral representation. We consider the graded microstructure as

$$\eta(r) = \begin{cases} 1+ar & 0 < r \le R, \\ 0 & r > R, \end{cases}$$

where R is the radius of the graded sphere. Thus, from Eq. (5.129) the dielectric constant in the graded sphere is given by

$$\varepsilon(r) = \varepsilon_2 \left(1 - \frac{\eta(r)}{s} \right). \tag{5.144}$$

In the dilute limit the effective dielectric constant of a small volume fraction p of graded spheres embedded in a host medium is given by [71, 148]

$$\varepsilon_e = \varepsilon_2 + 3\varepsilon_2 pb, \tag{5.145}$$

where b is the dipole factor of graded spheres embedded in a host as given in Ref. [12]. Using Eq. (5.134) and Eq. (5.145), the reduced response can be obtained as

$$F(s) = -3\varepsilon_2 pb. \tag{5.146}$$

Thus, the spectral density function of a graded sphere can be given through a numerical evaluation of Eq. (5.139).

B. Spectral density function of multilayer composites

A multilayer composite is a special case of graded composites. The gradation becomes continuous as the number of layers approaches infinity. To investigate the multilayer effect, we shall use a finite difference approximation for the graded profile (Eqs. (5.141) and (5.144)) for a finite number of layers. To mimic a multilayer system, we divide the interval $[0, 1]$ into N equally spaced sub-intervals, $[0, z_1], (z_1, z_2), \cdots, (z_N - 1, 1]$. Then we adopt the midpoint value of $\varepsilon(z)$ for each sub-interval as the dielectric constant of that sublayer. In this way, we calculate the effective dielectric constant, eigenvalues, as well spectral density function for each N. It is worth noting that the results of $N \rightarrow \infty$ (e. g., $N = 1024$ recovers the results of graded composites.

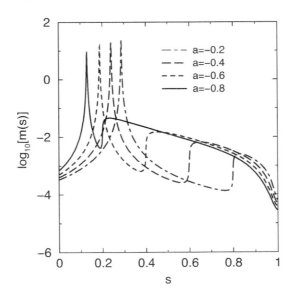

Figure 5.39. (Spectral density function of a graded sphere with volume fraction $p = 0.1$. After Ref. [57].

In addition to multilayer films, we can use the above approach to study the much simpler problem of a two-layer film. In this system, we have two layers of dielectric constants ε_1, ε_2, and host ε_0. Thickness are hy, $h(1-y)$, and $1-h$, respectively, where y is the length ratio between component ε_1 and component ε_2. We also define two microstructure parameters,

η_1 and η_2. If we let $s = 1/(1-\varepsilon_1/\varepsilon_0)$, then $\eta_1 = 1$, and $\eta_2 = (\varepsilon_0 - \varepsilon_2)/(\varepsilon_0 - \varepsilon_1)$. According to Eq. (5.141), the effective dielectric constant of the two-layer film is now given by

$$\frac{1}{\varepsilon_e} = \frac{hy}{\varepsilon_1} + \frac{h(1-y)}{\varepsilon_2} + \frac{1-h}{\varepsilon_0}.$$

According to Eq. (5.134), the reduced response can be given by

$$F(s) = \frac{F_1}{s - s_1} + \frac{F_2}{s - s_2}, \tag{5.147}$$

where

$$F_1 = \frac{h(s_1(y - y\eta + \eta) - \eta)}{s_1 - s_2}, \tag{5.148}$$

$$F_2 = -\frac{h(s_2(y - y\eta + \eta) - \eta)}{s_1 - s_2}, \tag{5.149}$$

$$s_1 = \frac{1}{2}[1 - h(y - y\eta + \eta) + \eta \tag{5.150}$$

$$-\sqrt{4\eta(-1+h) + (1 - h(y - y\eta + \eta) + \eta)^2}], \tag{5.151}$$

$$s_2 = \frac{1}{2}[1 - h(y - y\eta + \eta) + \eta \tag{5.152}$$

$$+\sqrt{4\eta(-1+h) + (1 - h(y - y\eta + \eta) + \eta)^2}]. \tag{5.153}$$

From the sums of F_1 and F_2 and the integral of graded microstructure $\eta(z)$ given by Eq. (5.140), we can check that the sum rule expressed by Eq. (5.136) is obeyed. It should also be noted that there are two poles in the expression for the reduced response corresponding to two peaks in the spectral density function. If $h = 1$, then $s_1 = 0$, that is, one peak is located at zero, which is explicitly shown in Fig. 5.38(a).

Similarly, we can also apply our graded spectral representation to a single-shell sphere of core dielectric constant ε_1, covered by a shell of ε_2, and suspended in a host of ε_0. In this example, we can also define two microstructure parameters η_1 and η_2. If we let $s = 1/(1 - \varepsilon_1/\varepsilon_0)$, then $\eta_1 = 1$, and $\eta_2 = (\varepsilon_0 - \varepsilon_2)/(\varepsilon_0 - \varepsilon_1)$. The dipole factor of single-shell sphere is given [71, 148]

$$b = \frac{\varepsilon_2 - \varepsilon_0 + (\varepsilon_0 + 2\varepsilon_2)xf^3}{\varepsilon_2 + 2\varepsilon_0 + 2(\varepsilon_2 - \varepsilon_0)xf^3},$$

where f is the ratio between radius core and radius shell, and x is given by

$$x = \frac{\varepsilon_1 - \varepsilon_2}{\varepsilon_1 + 2\varepsilon_2}.$$

Then, we can also write Eq. (5.146) similarly to Eq. (5.147), where the residues and

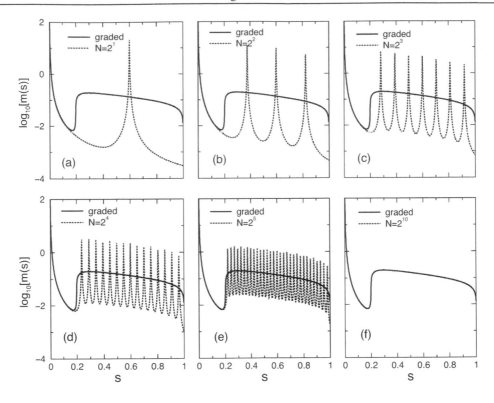

Figure 5.40. Spectral density function of various multilayer film with $\varepsilon_0 = 1$ and $a = -0.8$. After Ref. [57].

eigenvalues are given by,

$$F_1 = \frac{-3ps_1[(-1+\eta)y^3 - \eta] - \eta p[1 - 2(-1+\eta)y^3 + 2\eta]}{3(s_1 - s_2)}, \tag{5.154}$$

$$F_2 = \frac{3ps_2[(-1+\eta)y^3 - \eta] + \eta p[1 - 2(-1+\eta)y^3 + 2\eta]}{3(s_1 - s_2)}, \tag{5.155}$$

$$s_1 = \frac{1}{6}\left[1 + 3\eta - \sqrt{1 + (2 - 8y^3)\eta + (1 + 8y^3)\eta^2}\right], \tag{5.156}$$

$$s_2 = \frac{1}{6}\left[1 + 3\eta + \sqrt{1 + (2 - 8y^3)\eta + (1 + 8y^3)\eta^2}\right]. \tag{5.157}$$

Analysis shows that the spectral representation for $N = 2$ contains two simple poles corresponding to two peaks in the spectral density function. Therefore, we draw the conclusion that, N peaks are a result of N layers. Moreover, $N - 1$ peaks will accumulate into a continuous broad absorption spectrum when N tends toward infinity, which can be seen from Fig. 5.40(f) and 5.41(f).

We are now in a position to do some numerical calculations of the spectral density function from Eqs. (5.137) and (5.139). A small but finite imaginary part in the complex parameter has been used in the calculations. Without any loss of generality, we choose $L = 1$ and $R = 1$ for convenience. We show the effect of different graded profiles, as well as the effect of the thickness of the inclusion. It should be noted, that in all figures the range of s is limited to $[0, 1]$, because we chose $-1 < a < 0$ which limits the value of η into $[0, 1]$.

Fig. 5.37 displays the dielectric profile of a graded film (Fig. 5.37(a)) and a graded sphere (Fig. 5.37(b)). This figure obviously shows that the dielectric constant varies with the position in inclusion while a constant in host medium. Also, different values of a accord with different graded materials.

In Fig. 5.38(a), we plot the spectral density function $m(s)$ of a graded film without an interface against the spectral parameter for various graded microstructures $\eta(z)$. It is evident that there is always a broad continuous band in the spectral density function. Both the strength as well as the width of the continuous part of $m(s)$ increase with the gradient of the dielectric profile. Thus, the previous results of the broad surface-plasmon band can be expected. Note that there is a sharp peak at $s = 0$, which is also present in a homogeneous film. In Fig. 5.38(b), we plot the spectral density function of a graded film meeting a homogeneous medium at an interface for various graded microstructure $\eta(z)$. Again, there is always a broad continuous band in the spectral density function. However, the sharp peak has now shifted to a finite value of s, which is also present in a homogeneous film.

In Fig. 5.39, the spectral density function of graded sphere is displayed for a volume fraction $p = 0.1$. In this case, the interface always exists. It is clear that a broad continuous function in the spectral density function is always observed, as well as the shift of the sharp peak. However, the decrease of the broad continuous function is more abrupt for graded sphere than for graded film with increasing s.

Figures 5.40 and 5.41 display the spectral density function for a multilayer film and a sphere, respectively. It is clear that there are always N sharp peaks for N layers. Moreover, it is worth noting that there occurs a transition from sharp peaks to a broad continuous band with increasing N (see Fig. 5.40(f) and Fig. 5.41(f)), that is, the graded results are recovered by the limit results of $N \to \infty$. In particular, we had obtained the analytical expression of spectral density function for $N = 2$. There are two resonances corresponding to the two peaks in Fig. 5.40(a) and Fig. 5.41(a).

We have investigated a graded composite film and a sphere by means of the Bergman-Milton spectral representation. It has been shown that the spectral density function can be obtained analytically for a graded system. However, unlike in the case of homogeneous constituent components, the characteristic function is a continuous function due to the presence of gradation. Moreover, the derivation as well as some salient properties, namely, the sum rule, the definition of inner product, the definition of the integral-differential operators, and the range of spectral parameters, do change because of the continuous variation of the dielectric profile within the constituent components. It should be noted that in graded composite, the eigenvalues are not limited to $[0, 1]$, and they can be extended to $-\infty \leq s_n \leq \infty$ for the full region η, i.e., $-\infty \leq \eta \leq \infty$. Here, however for simplicity, we investigated the spectral density function in $0 \leq s \leq 1$ by choosing $-1 < a < 0$ to limit the value of η into $[0, 1]$.

We also study multilayer composites and calculated the spectral density function versus the number of layers, to explicitly demonstrate that the broad continuous spectrum arises from the accumulation of poles when the number of layers tends to infinity. This finding coincides with the broad surface-plasmon absorption band associated with the optical properties of graded composites.

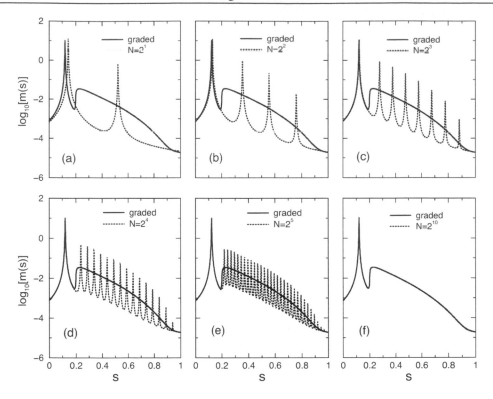

Figure 5.41. Spectral density function of various multi-shell sphere with $\varepsilon_0 = 1$, $a = -0.8$ and $p = 0.1$. After Ref. [57].

To sum up, we have investigated the spectral density function of graded film and graded sphere, as well as multilayer cases. There is always a broad continuous function in the spectral density function in graded composite, but simple poles in multilayer composite, the number of pole depends on the number of layers. Moreover, there is a gradual transition from sharp peaks to a broad continuous band until the graded composite results recover in the limit of $N \rightarrow \infty$.

Chapter 6

Magneto-Controlled Nonlinear Optical Materials

Ferrofluids (magnetic fluids) are colloidal suspensions containing single domain nanosize ferromagnetic particles dispersed in a carrier liquid (e.g., kerosene or water) [149]. Since these particles can easily interact via applied magnetic fields, which in turn can affect the viscosity and structural properties, ferrofluids possess a wide variety of potential applications in many fields ranging from mechanical engineering [150, 151] to biomedical applications [152, 153].

In this chapter, we shall include a nonlinear metallic shell in the field-responsive ferrofluids, and then tune the effective linear and nonlinear optical properties by applying a magnetic field. Such a proposed magneto-controlled ferrofluid-based nonlinear optical material can serve as optical materials which have anisotropic nonlinear optical properties and a giant enhancement of nonlinearity, as well as an attractive figure of merit. The approach may also be useful for presenting electro-controlled nonlinear optical materials. In so doing, one can add a nonlinear metallic shell to the nanoparticles in electrorheological nanofluids, too.

1. Third-Order Nonlinearity

We shall exploit theoretically a nonlinear optical material whose nonlinear optical properties and nonlinearity enhancement can be tuned by applying an external magnetic field - thus also called *magneto-controlled nonlinear optical materials* (Fig. 6.1). Devices that could benefit from these materials include optical switches, optical limiters, etc.

Ferromagnetic nanoparticles, typically consisting of magnetite or cobalt, have a typical diameter of 10 nm, and carry a permanent magnetic moment (e. g., of the strength $\sim 2.4 \times 10^4 \mu_B$ for magnetite nanoparticles, where μ_B denotes the Bohr magneton) [154]. As the ferromagnetic nanoparticles are suspended in a host fluid like water or kerosene, they can easily form particle chains under the application of external magnetic fields [154], thus yielding a magnetic-field-induced anisotropical structure. Recently, a non-magnetic golden shell was used to enhance the stability of the ferromagnetic nanoparticle against air and moisture [155]. Below we shall show that the effective nonlinear optical response of the suspension that contains ferromagnetic nanoparticles with metallic nonlinear shells can be enhanced significantly due to the effect of the magnetic-field-induced anisotropy.

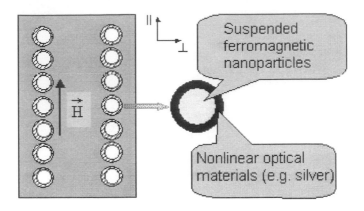

Figure 6.1. Design for a nonlinear optical materials, subjected to an external magnetic field **H**. ∥ (or ⊥): Longitudinal (or transverse) field cases corresponding to the fact that the E−field of an incident light is parallel (or perpendicular) to the nanoparticle chain.

The third-order nonlinear susceptibility χ_s of metallic (say typically, noble metals like gold and silver) shells is very large when compared to that of the magnetite or cobalt core and the host fluid like water. Let us start by considering ferromagnetic linear nanoparticles of linear dielectric constant ε_1'' coated with a non-magnetic metallic nonlinear shell of ε_1' and χ_s which are suspended in a linear host fluid of ε_2. That is, in the shells, there is a nonlinear relation between the displacement $\mathbf{D}_s$ and the electric field $\mathbf{E}_s$, $\mathbf{D}_s = \varepsilon_1' \mathbf{E}_s + \chi_s |\mathbf{E}_s|^2 \mathbf{E}_s$, where ε_1' is given by the Drude form, $\varepsilon_1' = 1 - \omega_p^2/[\omega(\omega + \gamma i)]$, where ω_p and γ stand for the plasmon frequency and the relaxation rate, respectively, and ω denotes the frequency of the incident light. In what follows, the thickness of the shell and the radius of the core are respectively denoted as d and R.

Next, we restrict our discussion to the quasi-static approximation, under which the structured particle or the whole suspension can be regarded as an effective homogeneous one. It is known that the effective third-order nonlinear susceptibility $\bar{\chi}$ of an area [here, the area represents the structured particle and the whole suspension, respectively, see Eqs. (6.7) and (6.10) below] is defined as [66, 71]

$$\bar{\chi} = \frac{1}{V|E_0|^2 E_0^2} \int_V \chi(\mathbf{r}) |\nabla \phi_0(\mathbf{r})|^2 [\nabla \phi_0(\mathbf{r})]^2 d\mathbf{r}, \qquad (6.1)$$

which is in terms of zeroth-order potential $\phi_0(\mathbf{r})$ only, see Eqs. (6.3-6.5) below. In Eq. (6.1) E_0 denotes the external applied electric field, V the volume of the area under consideration, $\mathbf{r}$ the local position inside the medium (r the distance from the particle center to the point of interest), and $\chi(\mathbf{r})$ an $\mathbf{r}$-dependent third-order nonlinear susceptibility. To obtain the effective nonlinear susceptibility of the structured particle which contains a linear core with a nonlinear shell, we should obtain the zeroth-order potentials which are actually obtained for the system in which the nonlinear characteristic of shells disappears, $\chi_s = 0$. Under the quasi-static approximation, the Maxwell equations read

$$\nabla \times \mathbf{E} = 0 \text{ and } \nabla \cdot \mathbf{D} = 0, \qquad (6.2)$$

and hence $\mathbf{E} = -\nabla\phi$, where ϕ is an electric potential. Solving Eqs. (6.2) [or the corresponding Laplace equation $\nabla^2\phi = 0$], we obtain the zeroth-order potentials for the core ϕ_0^c, the shell ϕ_0^s, and the host ϕ_0^h

$$\phi_0^c = -c_1 E_0 r \cos\theta, r < R, \tag{6.3}$$
$$\phi_0^s = -E_0(c_2 r - c_3 r^{-2})\cos\theta, R < r < R+d, \tag{6.4}$$
$$\phi_0^h = -E_0(r - c_4 r^{-2})\cos\theta, r > R+d, \tag{6.5}$$

where θ is the angle between the external field and the line joining the particle center and the point under investigation, and the coefficients c_1, c_2, c_3, and c_4 are determined by the appropriate boundary conditions. Owing to Eq. (6.1), *the effective third-order nonlinear susceptibility of the structured particle χ_1 can be given by*

$$\chi_1 \frac{\langle |\nabla\phi_0(\mathbf{r})|^2 [\nabla\phi_0(\mathbf{r})]^2 \rangle_{r \leq R+d}}{|E_0|^2 E_0^2} = f\chi_s \frac{\langle |\nabla\phi_0(\mathbf{r})|^2 [\nabla\phi_0(\mathbf{r})]^2 \rangle_{R < r \leq R+d}}{|E_0|^2 E_0^2}, \tag{6.6}$$

where f is the volume ratio of the shell to the core. Thus, we obtain

$$\chi_1 = \chi_s \frac{\beta}{\beta'}, \tag{6.7}$$

where $\beta = (3/5)[1/(1-f)^{1/3} - 1]|z|^2 z^2 (5 + 18x^2 + 18|x|^2 + 4x^3 + 12x|x|^2 + 24|x|^2 x^2)$ and

$$\beta' = \left| \frac{\varepsilon_2}{\varepsilon_2 + (\alpha/3)(\varepsilon_1 - \varepsilon_2)} \right|^2 \left(\frac{\varepsilon_2}{\varepsilon_2 + (\alpha/3)(\varepsilon_1 - \varepsilon_2)} \right)^2 \tag{6.8}$$

with $x = (\varepsilon_1'' - \varepsilon_1')/(\varepsilon_1'' + 2\varepsilon_1')$ and $z = (1/3)[\varepsilon_2(\varepsilon_1'' + 2\varepsilon_1')]/\{\varepsilon_1'[\varepsilon_2 + (\alpha/3)(\varepsilon_1'' - \varepsilon_2)]\}$. In Eq. (6.8), the effective linear dielectric constant ε_1 of each structured particle can be determined by the well-known Maxwell-Garnett formula with a high degree of accuracy

$$\frac{\varepsilon_1 - \varepsilon_1'}{\varepsilon_1 + 2\varepsilon_1'} = (1-f)\frac{\varepsilon_1'' - \varepsilon_1'}{\varepsilon_1'' + 2\varepsilon_1'}. \tag{6.9}$$

It is worth noting that for the above derivation a local magnetic-field factor α has been introduced, see Eq. (6.8). In detail, α denotes the local field factors α_L and α_T for longitudinal and transverse field cases, respectively. Here the longitudinal (or transverse) field case corresponds to the fact that the E-field of the light is parallel (or perpendicular) to the particle chain. Similar factors in electrorheological fluids were measured by using computer simulations [156, 157], and obtained theoretically [40, 158] according to the Ewald-Kornfeld formulation. There is a sum rule for α_L and α_T, $\alpha_L + 2\alpha_T = 3$ [41]. The parameter α measures the degree of anisotropy, which is induced by the applied magnetic field H. More precisely, the degree of the field-induced anisotropy is measured by how much α deviates from unity, $1 < \alpha_T < 3$ for transverse field cases and $0 < \alpha_L < 1$ for longitudinal field cases. As H increases α_T and α_L should tend to 3 and 0, respectively, which is indicative of the formation of more and more particle chains as evident in experiments [154]. So, a crude estimate of α can be obtained from the contribution of chains [159], namely, $\alpha = [4\pi(d+R)^3/p]\sum_{n=1}^{\infty} n\gamma_n(H)g_n$, where p denotes the volume fraction of the structured particles in the suspension, g_n the depolarization factor for a chain with n structured particles, and $\gamma_n(H)$ the density of the chain which is a function of H. It is noteworthy that for given p $\gamma_n(H)$ also depends on the dipolar coupling constant which relates the dipole-dipole interaction energy of two contacting particles to the thermal energy. Now, the system of interest can be equivalent to the one in which all the particles with linear dielectric constant ε_1 [Eq. (6.9)] and nonlinear susceptibility χ_1 [Eq. (6.7)] are embedded in a host fluid with ε_2. For the equivalent system, it is easy to solve the corresponding Maxwell equations [Eqs. (6.2)], in order to get the zeroth-order potentials in the particles and the host. According to Eq. (6.1), we obtain *the effective third-order nonlinear susceptibility of the whole suspension* χ_e as $\chi_e = p\chi_1\beta'$, which can be rewritten as

$$\chi_e = p\chi_s\beta, \qquad (6.10)$$

The substitution of $\alpha = 1.0$ (i.e., the isotropic limit) into Eq. (6.10) yields the same expression as derived in Ref. [71] in which the dielectric constants of the core and shell of structured particles were, however, assumed to be real rather than complex. On the other hand, the effective linear dielectric constant of the whole suspension under present consideration ε_e can be given by the developed Maxwell-Garnett approximation which works for suspensions with field-induced anisotropic structures [40]

$$\frac{\varepsilon_e - \varepsilon_2}{\alpha\varepsilon_e + (3-\alpha)\varepsilon_2} = p\frac{\varepsilon_1 - \varepsilon_2}{\varepsilon_1 + 2\varepsilon_2}. \qquad (6.11)$$

For numerical calculations, without loss of generality we take $f = 0.65$, $p = 0.2$, $\varepsilon_1'' = -25 + 4i$, $\varepsilon_2 = 1.77$ (high-frequency limit dielectric constant of water), and $\gamma = 0.01\omega_p$. We further see χ_s to be a real and positive frequency-independent constant, in order to focus on the nonlinearity enhancement. Figures 6.2 and 6.3 display the linear optical absorption $\mathrm{Im}(\varepsilon_e)$, the enhancement of the third-order optical nonlinearity $|\chi_e|/\chi_s$, and the FOM (figure of merit) $|\chi_e|/[\chi_s\mathrm{Im}(\varepsilon_e)]$, as a function of normalized frequency ω/ω_p, for (Fig. 6.2) longitudinal and (Fig. 6.3) transverse field cases. Here the frequency ω is normalized by ω_p (rather than a specific value of ω_p), so that the result could be valid for general cases. As mentioned before, $\alpha = 1.0$ corresponds to the isotropic limit. In this case, there is no external magnetic field, and hence all the structured particles are randomly suspended. The figures show that the existence of nonlinear shells causes an enhancement of nonlinearity to appear, see Fig. 6.2(b) and Fig. 6.3(b), thus yielding a large FOM, see Fig. 6.2(c) and Fig. 6.3(c). Such a nonlinearity enhancement induced by shell effects was already reported [71]. The main feature of Figures 6.2 and 6.3 is the effects of external magnetic fields. As α_L changes from 1.0, to 0.6, and to 0.2, (namely, as α_T varies from 1.0, to 1.2, and to 1.4) the external magnetic field is adjusted from zero, to low strength, and to high strength. Due to the interaction between the ferromagnetic nanoparticles and the magnetic field, more and more particle chains are caused to appear naturally, thus yielding a magnetic-field-induced anisotropic structure in the suspension. It is evident to observe that the plasmon peak is caused to be blue-shifted for longitudinal field cases as the magnetic field increases. However, for transverse field cases, the plasmon peak displays a red-shift for the increasing magnetic field. In other words, the optical absorption is induced to be anisotropic due to the application of the external magnetic field which produces an

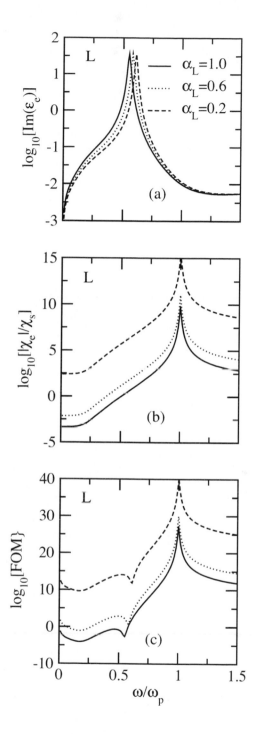

Figure 6.2. (a) The linear optical absorption $\mathrm{Im}(\varepsilon_e)$, (b) the enhancement of the third-order optical nonlinearity $|\chi_e|/\chi_s$, and (c) the FOM $|\chi_e|/[\chi_s \mathrm{Im}(\varepsilon_e)]$ versus the normalized incident angular frequency ω/ω_p, for various strengths of the external magnetic field which are represented by local-field factors α_L, for longitudinal field cases (L). After Ref. [7].

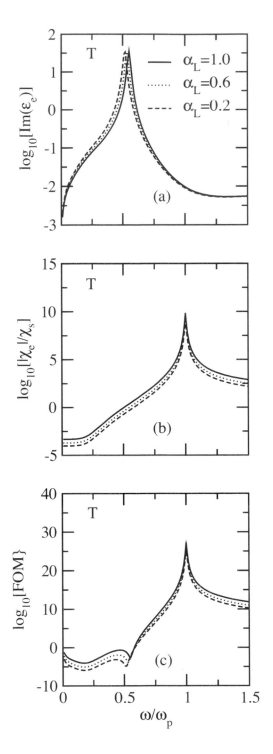

Figure 6.3. Same as Fig. 6.2, but for transverse field cases (T). After Ref. [7].

anisotropic structure. In fact, the optical absorption arises from the surface plasmon resonance, which is obtained from the imaginary part of the effective dielectric constant. For single metallic particles in the dilute limit, it is well known that there is a large absorption when the resonant condition $\varepsilon_1' + 2\varepsilon_2 = 0$ is fulfilled. When there is a larger volume fraction p of structured particles and an anisotropy α of the suspension, the effective dielectric constant should be obtained from Eq. (6.11), thus yielding a modified resonant condition $(1 - p\alpha)\varepsilon_1 + (2 + p\alpha)\varepsilon_2 = 0$. So, the resonant frequency becomes larger (smaller) than the isotropic limit ($\alpha = 1$) when α becomes smaller (larger) than 1. In other words, there is a blue (red) shift for the longitudinal (transversal) field cases. More interestingly, for longitudinal field cases, a giant enhancement of nonlinearity is shown as the magnetic field increases, see Fig. 6.2(b). In detail, the nonlinearity enhancement of a high-field case (say, $\alpha = 0.2$) can be of five orders of magnitude larger than that of the zero-field case ($\alpha = 1.0$). Inversely, a reduction of nonlinearity is found for transverse field cases, see Fig. 6.3(b). The magnitude of the nonlinearity reduction is very small in the transverse field case, when compared to that of the nonlinearity enhancement in the longitudinal field case. Owing to the giant enhancement of nonlinearity [see Fig. 6.2(b)], the FOM becomes much more attractive for longitudinal field cases [see Fig. 6.2(c)]. The FOM of a high-field case (say, $\alpha = 0.2$) can even be ten-order-of-magnitude enhanced in the longitudinal field case. However, the effect of the magnetic field on the FOM for transverse field cases seems to be uninteresting since the FOM is caused to be decreased slightly due to the nonlinearity reduction shown in Fig. 6.3(b). Since the permanent magnetic moment of the magnetite nanoparticles m is approximately $2.4 \times 10^4 \mu_B$ [154], we can estimate the threshold magnetic field $H_c = 14.3 \, \text{kA/m}$ (or threshold magnetic induction $B_c = 0.018 \, \text{T}$) above which the corresponding magnetic energy can overcome the thermal energy $1/40 \, \text{eV}$ so as to obtain appreciable anisotropy. Besides the magnetic energy, we should also compare the interaction energy. For instance, for two touching magnetite nanoparticles, the interaction between them is proportional to $m^2/[2(d+R)]^3$, assuming the two structured particles to be in a head-to-tail alignment. Since the magnetic moment m goes as $(2R)^3$, the interaction energy could vary as $[2R^2/(d+R)]^3$. In order to break up the two touching nanoparticles, the thermal energy should be larger than the interaction energy. So, threshold field $H_c = 14.3 \, \text{kA/m}$ serves as an upper estimate. Nevertheless, for cobalt nanoparticles, the threshold field H_c should be lower due to larger permanent magnetic moments. To sum up, by including a metallic nonlinear shell in the system, one can tune the linear and nonlinear optical properties by applying a magnetic field. Such a proposed magneto-controlled nonlinear optical material can serve as optical materials which have anisotropic nonlinear optical properties and a giant enhancement of nonlinearity, as well as an attractive FOM.

2. Second-Harmonic Generation

Theoretical [160] and experimental [161] reports suggested that spherical particles exhibit a rather unexpected and nontrivial behavior, second-harmonic generation (SHG), due to the broken inversion symmetry at particle surfaces, despite their central symmetry which seemingly prohibits second-order nonlinear effects. In colloidal suspensions, the SHG response for centrosymmetric particles was experimentally reported [161]. Most recently, the SHG from centrosymmetrical structure has received an extensive attention (e.g., see Refs. [162, 163]). In order to obtain a second-harmonic generation (SHG) with magnetic-field controllabilities, in view of recent advancements in the fabrication of nanoshells [38, 39] and single domain ferromagnetic nanoparticles [164], we shall theoretically suggest a nonlinear optical material in which single domain ferromagnetic nanoparticles coated by a nonmagnetic nanoshell with an intrinsic SHG susceptibility are suspended in a nonmagnetic host fluid (Fig. 6.1). For such a material, there is not only an incident light, but also an external magnetic field **H**. The latter yields the formation of chains of coated nanoparticles [154], thus changing the microstructure of the material. Similar to Section 1., this also produces magnetic-field-controllable SHG responses. We shall present a model to show and understand the physics/mechanism of the SHG. This SHG is expected to receive a broad interest in the physics, optics, and engineering communities, because it is difficult or impossible to achieve with conventional, naturally occurring materials or random composites [7, 14, 16, 37, 73, 75, 77, 80, 127, 165].

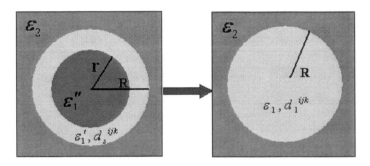

Figure 6.4.   Schematic graph showing the equivalence between a coated inhomogeneous nanoparticle (left) and a homogeneous nanoparticle (right) according to Eqs. (6.12) and (6.13).

Similarly, let us consider a linear ferromagnetic spherical nanoparticle with dielectric constant ε_1'' and radius r, which is coated by a nonlinear optical nonmagnetic nanoshell (e.g., noble metals like silver or gold) with frequency-dependent dielectric constant $\varepsilon_1'(\omega)$ and intrinsic SHG susceptibility $d_s^{ijk}(-2\omega;\omega,\omega)$ with each of the superscripts running over the three Cartesian indices. Here ω denotes the angular frequency of a monochromatic external electric field, and the radius of the whole coated nanoparticle is represented as R in the following. All the coated nanoparticles are suspended in a linear nonmagnetic host fluid of ε_2. In the nanoshell, the local constitutive relation between the displacement field $\mathbf{D}_s$ and the electric field $\mathbf{E}_s$ in the static case is given by, $D_s^i = \sum_j \varepsilon_1'(\omega)^{ij} E_s^j + \sum_{jk} d_s^{ijk}(-2\omega;\omega,\omega) E_s^j E_s^k$ ($i = x, y, z$), where D_s^i and E_s^i are the ith component of $\mathbf{D}_s$ and $\mathbf{E}_s$, respectively. Here $\varepsilon_1'(\omega)^{ij} = \varepsilon_1'(\omega)\delta_{ij}$ denotes the linear dielectric constant, which is assumed for simplicity to be isotropic. Upon certain symmetry, one can have $d_s^{iii}(-2\omega;\omega,\omega) \neq 0$ ($i = x, y, z$) as $d_s^{xxx}(-2\omega;\omega,\omega) = d_s^{yyy}(-2\omega;\omega,\omega) = d_s^{zzz}(-2\omega;\omega,\omega)$. If a monochromatic external field is applied, the nonlinearity in the system will generally generate local potentials and fields at all harmonic frequencies. For a finite-frequency external electric field of the form $E_0 = E_0(\omega)e^{-i\omega t} + c.c.$, the equivalent and effective SHG susceptibilities for the coated nanoparticle and the whole suspension, $d_1^{zzz}(-2\omega;\omega,\omega)$ [Eq. (6.13)] and $d_e^{zzz}(-2\omega;\omega,\omega)$ [Eq. (6.15)], can be extracted by considering the volume average of the displacement field at the frequency 2ω in the inhomogeneous medium. The electric field $\mathbf{E}_s$ in the nanoshell can be calculated [77] using standard electrostatics, by solving the corresponding Maxwell equation $\nabla \times \mathbf{E}_s = 0$, which implies that $\mathbf{E}_s = -\nabla \phi$, where ϕ is an electric potential. Next, the equivalent linear dielectric constant $\varepsilon_1(\omega)$ for the coated nanoparticle (Fig. 6.4) can be given by the Maxwell-Garnett formula [80]

$$\frac{\varepsilon_1(\omega) - \varepsilon_1'(\omega)}{\varepsilon_1(\omega) + 2\varepsilon_1'(\omega)} = (1 - f)\frac{\varepsilon_1'' - \varepsilon_1'(\omega)}{\varepsilon_1'' + 2\varepsilon_1'(\omega)}, \tag{6.12}$$

where $f = 1 - r^3/R^3$ is the volume ratio of the nanoshell to the whole coated nanoparticle. The Maxwell-Garnett formula is a well-known asymmetrical effective medium theory, and may thus be valid for a low concentration of nanoparticles in composites [70]. While treating a single coated nanoparticle with full range $0 \le f \le 1$, the Maxwell-Garnett formula [Eq. (6.12)] holds for the calculation of $\varepsilon_1(\omega)$ indeed, due to the natural existence of asymmetry in the coated nanoparticle. The solution of $\mathbf{E}_s$ [77] can be used to derive the equivalent SHG susceptibility for the coated nanoparticle (Fig. 6.4), $d_1^{iii}(-2\omega; \omega, \omega)$,

$$d_1^{iii}(-2\omega; \omega, \omega) = f d_s^{iii}(-2\omega; \omega, \omega) \sum_{j=x}^{z} \left\langle \frac{E_{s,j}(2\omega)}{E_{0,i}(2\omega)} \left(\frac{E_{s,j}(\omega)}{E_{0,i}(\omega)}\right)^2 \right\rangle_s, \tag{6.13}$$

where $\langle \cdots \rangle_s$ denotes a volume average over the nanoshell. To show the feature of the proposed material, we assume the optical responses [namely, $\varepsilon_1(\omega)$ and $d_1^{iii}(-2\omega; \omega, \omega)$] of an equivalent spheroid or a chain (see Fig. 6.5) to be the same as those of each coated nanoparticle inside the spheroid or chain. For convenience, the suspension is further assumed to be the one that contains identical equivalent spheroids with geometrical depolarization factor $\alpha_{\parallel}$ (or $\alpha_{\perp}$) along major (or minor) axis (Fig. 6.5). In the following, α is also called *local magnetic field factors*, because, from the physical point of view, the spheroids (or chains) are just formed due to the application of external magnetic fields. In this connection, the summation term in Eq. (6.13) admits $\{\Pi(2\omega)\Pi^2(\omega) + (4/5)(rR)^{-3}[\Pi(2\omega)p_s^2(\omega) + 2\Pi(\omega)p_s(2\omega)p_s(\omega)] + (8/35)(r^3 + R^3)/(r^6R^6)p_s(2\omega)p_s^2(\omega)\}/[E_{0,i}(2\omega)E_{0,i}^2(\omega)]$, where $\Pi(\omega) = T_s(\omega)E_{0,i}(\omega)$ and $p_s(\omega) = b_s(\omega)r^3T_s(\omega)E_{0,i}(\omega)$ with $b_s(\omega) = (\varepsilon_1'' - \varepsilon_1'(\omega))/(\varepsilon_1'' + 2\varepsilon_1'(\omega))$ and $T_s(\omega) = [\Theta(\omega) + 2b_s(\omega)(\Theta(\omega) - 1)(1 - f)]^{-1}$. Here $\Theta(\omega) = [\varepsilon_2 + \alpha(\varepsilon_1'(\omega) - \varepsilon_2)]/\varepsilon_2$. Similarly, in the derivation, we have also introduced the local magnetic field factor α that denotes the local magnetic field factors $\alpha_{\parallel}$ and $\alpha_{\perp}$ for longitudinal and transverse field cases, respectively (Figs. 6.1 and 6.5). Due to different scale in use, there is a relation between the $\alpha_{\parallel}$ (or $\alpha_{\perp}$) and the α_L (or α_T) in Section 1., namely, $\alpha_{\parallel} = \alpha_L/3$ (or $\alpha_{\perp} = \alpha_T/3$). There is a sum rule for $\alpha_{\parallel}$ and $\alpha_{\perp}$, $\alpha_{\parallel} + 2\alpha_{\perp} = 1$ [41]. Again, the parameter α measures the degree of structural anisotropy due to the formation of nanoparticle chains, which is induced to appear by the external magnetic field $\mathbf{H}$. More precisely, the degree of the magnetic-field-induced anisotropy is measured by how much α deviates from $1/3$, $1/3 < \alpha_{\perp} < 1$ and $0 < \alpha_{\parallel} < 1/3$. As H increases, $\alpha_{\perp}$ and $\alpha_{\parallel}$ should tend to 1 and 0, respectively, which is indicative of the formation of longer nanoparticle chains (or equivalent spheroids). Therefore, α should be a function of external magnetic fields H. Specifically, for $H = 0$ there is $\alpha_{\parallel} = \alpha_{\perp} = 1/3$, which corresponds to an isotropic system in which all the coated nanoparticles are randomly distributed in the suspension. The substitution of $\alpha = 1/3$ into

Eq. (6.13) yields the same expression as the Eq. (15) in Ref. [77] in which a random composite of particles with nonlinear non-metallic shells was investigated. Alternatively, according to the calculation of major-axis depolarization factor L of prolate spheroids [41], $L = 1/(1-\rho^2) + \rho/(\rho^2-1)^{3/2} \ln(\rho + \sqrt{\rho^2-1})$, where $\rho(>1)$ is the ratio between the major and minor axes of the elliptic cross section, α can be given in terms of the number n of nanoparticles in an equivalent spheroid (or a chain)

$$\alpha_{\parallel} = \frac{1}{1-n^2} + \frac{n}{(n^2-1)^{3/2}} \ln(n + \sqrt{n^2-1}). \qquad (6.14)$$

Throughout the section, while both the nano-shell and host fluid are nonmagnetic, the core is ferromagnetic. The existence of ferromagnetism in the core makes the chain formation possible, as long as an external magnetic field H is applied [154]. We have added the magnetic contribution to the expressions for optical responses through the local magnetic field factor α. So far the exact relation between α and H lacks because it relates to complicated suspension hydrodynamics and kinetics at nonequilibrium. Nevertheless, the results obtained from Fig. 6.6 are valid for equilibrium systems in which neither hydrodynamics nor kinetics can affect the SHG. Without loss of any generality, to capture the features and their physics of the proposed material, in Fig. 6.6 we use α to represent the strength of the external magnetic field H.

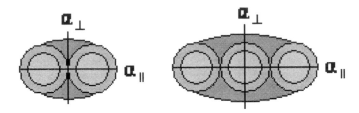

Figure 6.5. Schematic graph showing the equivalence between nanoparticle chains and spheroids with geometrical major-axis (or minor-axis) depolarization factor $\alpha_{\parallel}$ (or $\alpha_{\perp}$) [Eq. (6.14)]. The major axis is parallel to external magnetic fields. α is also called local magnetic field factors.

Now we see the suspension as the one in which the equivalent spheroids with $\varepsilon_1(\omega)$ [Eq. (6.12)] and $d_1^{iii}(2\omega;\omega,\omega)$ [Eq. (6.13)] are embedded in the host fluid. Owing to the $z-$directed external magnetic field, all the spheroids should also be directed along z axis, but with the locations being randomly distributed. According to the general expression for the effective SHG susceptibility [75, 77], we take one step forward to express the effective SHG susceptibility $d_e^{iii}(-2\omega;\omega,\omega)$ for the whole suspension in the dilute limit,

$$d_e^{iii}(-2\omega;\omega,\omega) = p d_1^{iii}(-2\omega;\omega,\omega)\Gamma(2\omega)\Gamma^2(\omega), \qquad (6.15)$$

where p is the volume fraction of the coated nanoparticles. In Eq. (6.15), $\Gamma(\omega)$ is a local electric field enhancement factor, and it is obtained by deriving the factor in a spheroid of depolarization factor α with principle axes along external electric fields, $\Gamma(\omega) = \varepsilon_2/[\varepsilon_2 + \alpha(\varepsilon_1(\omega) - \varepsilon_2)]$.

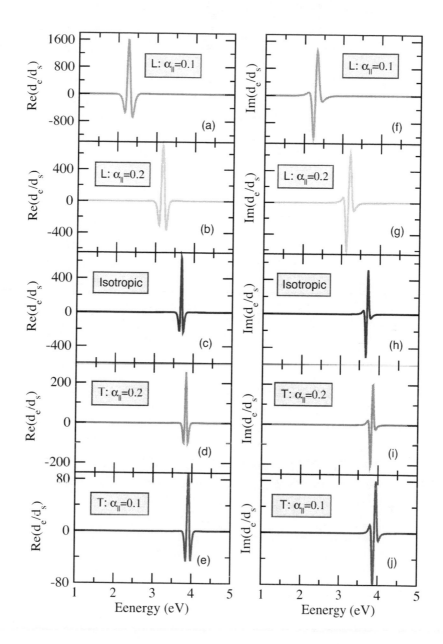

Figure 6.6. The [(a)-(e)] real and [(f)-(j)] imaginary parts of the effective SHG suscepti-bility $d_e^{iii}(-2\omega;\omega,\omega) \equiv d_e$ normalized by the intrinsic SHG susceptibility in the nonlinear nanoshell $d_s^{iii}(-2\omega;\omega,\omega) \equiv d_s$, for different external magnetic fields represented by local magnetic field factors $\alpha_{\parallel}$, versus the energy of an incident light. Here L and T denote the longitudinal and transverse field cases, respectively. According to Eq. (6.14), the number n of nanoparticles in the chains is $n \approx 3$ for $\alpha_{\parallel} = 0.1$ and $n \approx 2$ for $\alpha_{\parallel} = 0.2$. Note in (d), (e), (i) and (j) the corresponding $\alpha_{\perp}$ in use can be calculated according to the relation $\alpha_{\parallel} + 2\alpha_{\perp} = 1$.

Since metal surfaces were used to obtain enhanced SHG responses [166], for our numerical simulations we take a Drude dielectric function (that is valid for noble metals within the frequency range of interest) for $\varepsilon_1'(\omega)$, $\varepsilon_1'(\omega) = \varepsilon(\infty) - (\varepsilon(0) - \varepsilon(\infty))\omega_p^2/[\omega(\omega + i\gamma)]$, where ω_p is the resonant plasmon frequency, $\varepsilon(\infty)$ the high-frequency limit dielectric constant, $\varepsilon(0)$ the static dielectric constant, and γ the collision frequency. Specifically, for silver, $\varepsilon(\infty) = 5.45$, $\varepsilon(0) = 6.18$, and $\omega_p = 1.72 \times 10^{16}$ rad/s [167]. In addition, we take $\gamma = 0.01\omega_p$ (a typical value for metals), $r = 5$ nm (a typical value for single domain ferromagnetic nanoparticles [154]), the thickness of nanoshells 1.9 nm (or $R = 6.9$ nm), $\varepsilon_1'' = -25 + 4i$ (e.g., for cobalt), frequency-independent dielectric constant $\varepsilon_2 = 1.77$ (high-frequency limit dielectric constant of water), and $p = 0.18$. Based on the values of r, R and p, we obtain the volume fractions, p_c and p_s, of the ferromagnetic and nonlinear optical components in the whole suspension, $p_c = 0.07$ and $p_s = 0.11$, according to the relations $R/r = (1 + p_s/p_c)^{1/3}$ and $p = p_c + p_s$. We shall investigate the light energy $1 \sim 5$ eV, which corresponds to the wavelength $\lambda = 248 \sim 1242$ nm, or the frequency range $\omega = 1.52 \times 10^{15} \sim 7.59 \times 10^{15}$ rad/s.

We show the effective SHG susceptibility $d_e^{iii}(-2\omega; \omega, \omega)$ of the whole suspensions in Fig. 6.6. For longitudinal field cases [Fig. 6.6(a), (b), (f), and (g)], as $\alpha_\parallel$ decreases (i.e., external applied magnetic field H increases, and longer nanoparticle chains are formed accordingly), the resonant peak in the SHG response is not only red-shifted (namely, located at a lower frequency), but also further enhanced, when compared to the isotropic case at zero external magnetic field $H = 0$ [Fig. 6.6(c) and (h)]. However, inverse behavior appears for transverse field cases [Fig. 6.6(d), (e), (i), and (j)]. In detail, for transverse field cases, as the external magnetic field increases, the resonant peak in the SHG signal is both reduced and blue-shifted (i.e., located at higher frequency), and hence becomes less attractive. The formation of nanoparticle chains due to the application of external magnetic fields changes the surrounding circumstance of each coated nanoparticle naturally, which in turn affects the local electric field in the nanoshells and hence shifts the resonant plasmon frequency at which the resonant peak appears (Fig. 6.6). In the presence of an external magnetic field, the SHG response becomes anisotropic (i.e., its strength in the longitudinal field differs from that in the transverse field), and the degree of anisotropy can further be adjusted by tuning the external magnetic field. Moreover, in longitudinal field cases the enhanced peak is red-shifted as longer nanoparticle chains are formed, i.e., the external magnetic field is increased. In the mean time, an enhancement of the SHG peak is achieved.

Since the permanent magnetic moment of magnetite nanoparticles m is approximately $2.4 \times 10^4 \mu_B$ (where μ_B denotes the Bohr magneton) [154], we can estimate the threshold magnetic field $H_c = 14.3$ kA/m (or threshold magnetic induction $B_c = 0.018$ T) above which the corresponding magnetic energy can overcome the thermal energy $1/40$ eV so as to obtain appreciable anisotropy. It is worth noting that $B_c \gg B_{c0}(\sim 10^{-4}$ T) where B_{c0} means the earth's magnetic induction. Besides the magnetic energy, we should also consider the interaction energy. In order to break up the touching nanoparticles, the thermal energy should be larger than the interaction energy. So, threshold field $H_c = 14.3$ kA/m serves as an upper estimate, whereas for cobalt nanoparticles the threshold field H_c should be lower due to larger permanent magnetic moments inside them.

In looking for experimental evidence of the magnetic-field controllabilities presented here, we note that Du and Luo have reported nonlinear optical effects in suspensions of ferromagnetic nanoparticles (with mean diameter 9 nm) in kerosene [168]. They observed that the nonlinear optical effect is enhanced by applying a moderate magnetic field. However, this mechanism was unclear at that time, as they claimed. Based on the present analysis, it seems not difficult to conclude that this enhanced nonlinear optical effect results from the magnetic field-induced anisotropic structure, thus being magnetic-field controllable. Owing to the pioneering experiment by Du and Luo [168], experimental tests of our model should be possible. In so doing, the key point is that the measurements should be done at equilibrium. This is because as an external applied magnetic field changes, the initial alignment of suspended nanoparticles into chains are always unstable (at nonequilibrium). Note our results obtained from Fig. 6.6 are valid for equilibrium systems in which neither hydrodynamics nor kinetics can affect the SHG. (For a non-equilibrium system, complex hydrodynamics and kinetics should be taken into account.) In this Chapter we have not included the effect of distributions of the number of nanoparticles in chains. Nevertheless, based on the same physics/mechanism, it is not difficult to understand that the SHG signal can become broad (not as sharp as in Fig. 6.6) once the distribution is taken into account. In summary, we have suggested a nonlinear optical material, and discovered the SHG with the magnetic-field controllability (i.e., magnetic-field-controllable anisotropy, red-shift and enhancement). The proposed material is expected to be valuable for optical applications like optical limiters, optical switches, etc.

Chapter 7

Electrorheological Nanofluids or Ferrofluids

Ferrofluids (magnetic fluids) are colloidal suspensions containing single domain nanosize ferromagnetic particles dispersed in a carrier liquid (e.g., kerosene or water) [149]. These particles are usually stabilized against agglomeration by coating them with long-chain molecules (steric stabilization) or decorating them with charged groups (electrostatic stabilization). Since these particles can easily interact via applied magnetic fields, which in turn can affect the viscosity and structural properties, ferrofluids possess a wide variety of potential applications in many fields ranging from mechanical engineering [150, 151] to biomedical applications [152, 153].

Electrorheology denotes the control of a material's flow properties (rheology) through an electric field [169, 170]. Giant electrorheological fluids [171] contain polarizable nanoparticles [e.g., $BaTiO(C_2O_4)_2$] coated with urea, embedded in a host fluid (e.g., silicone oil). When an electrorheological fluid [170, 171, 172, 173, 174, 175, 176, 177] is subjected to a strong external field, elongated chains or columns of polarizable dielectric particles (e.g., titanium particles) form immediately parallel to the field due to the anisotropic long-range particle interaction inside the liquid carrier (e.g., silicone or corn oil). Because of this sort of rapid field-induced aggregation, recently electrorheological fluids have received much attention [170, 171, 173, 174, 175, 176, 177] in both scientific research and industrial applications. For instance, electrorheological fluids were also proposed as a method of constructing shock absorbers on magnetically levitated trains.

In fact, giant electrorheological fluids belongs to a class of electrorheological nanofluids. This is because the size of the suspended particles in giant electrorheological fluids are generally of nano-scale. In other words, the common point for both ferrofluids and electrorheological nanofluids is that all the suspended particles are nanosized.

The authors [178] experimentally reported on the new observation of fully reversible light-induced cluster formation in sterically stabilized kerosene-based magnetite (Fe_3O_4) ferrofluids. When illuminated with visible light within their optical absorption band, μm-sized clusters are formed that contain of the order of 10^7 individual colloidal particles. After switching the illumination off, the attractive forces disappear and the clusters dissolve on a time scale of 1 min. Linear chains of several $100\,\mu$m length are formed upon illumination in the presence of a weak magnetic field (1 mT). To some extent, this new effect may be useful for patterning and optically triggered reversible assembly of controlled nano- and microstructures.

By using a theoretical approach based on thermodynamics, we shall present a different method to obtain nonlinear optical responses from electrorheological nanofluids or ferrofluids in which the components possess inherent *linear* responses only (namely, linear electrorheological fluids or ferrofluids), under the influence of light-striction. Thus, such linear suspensions can also serve as a nonlinear optical material because the effective third-order nonlinear optical susceptibility can be induced due to the light-striction effect. For clarity, it is worth noting that the above-mentioned linear suspensions represent the suspensions whose dielectric constants are independent of the external electric field. For the sake of convenience, below we shall only take electrorheological nanofluids as an example. The same footing will naturally hold for ferrofluids.

For investigating the light-striction effect, take the experimental situation as follows: There is a light-affected volume with volume V_c, in which the electric field and the dielectric displacement are denoted by E_c and D_c, respectively. Both of them should satisfy the usual electrostatic equations, namely

$$\nabla \cdot \mathbf{D}_c = 0, \tag{7.1}$$

$$\nabla \times \mathbf{E}_c = 0. \tag{7.2}$$

Here Eq. (7.2) implies that the electric field $\mathbf{E}_c$ can be expressed as the gradient of a potential ϕ,

$$\mathbf{E}_c = -\nabla\phi. \tag{7.3}$$

Under the appropriate boundary condition, the electrorheological nanofluid within the light-affected volume can be represented as a region of volume V_c, surrounded by surface S_s. Such kind of boundary condition is

$$\phi = -\mathbf{E} \cdot \mathbf{R} \text{ on } S_s, \tag{7.4}$$

which, if the electrorheological nanofluid within V_c were uniform, would give rise to an electric field which is identical to $\mathbf{E}$ everywhere within V_c. As a matter of fact, even in an inhomogeneous electrorheological nanofluid with this boundary condition, the volume average of the electric field $\langle \mathbf{E}_c \rangle$ within V_c still equals that of the external field $\langle \mathbf{E} \rangle$, i.e.

$$\langle \mathbf{E_c} \rangle = \frac{1}{V_c} \int \mathbf{E}_c(\mathbf{R}) \mathrm{d}^3 r = \langle \mathbf{E} \rangle. \tag{7.5}$$

It is worth noting that in this case there is no incident light outside the light-affected volume. Also, the whole electrorheological nanofluid with volume $V(> V_c)$ is situated at a constant pressure p.

The effective linear dielectric constant ε_e and effective third-order nonlinear susceptibility χ for the electrorheological nanofluid inside the light-affected volume are defined as

$$\langle \mathbf{D}_c \rangle = \varepsilon_e \langle \mathbf{E} \rangle + 4\pi\chi |\langle \mathbf{E} \rangle|^2 \langle \mathbf{E} \rangle, \tag{7.6}$$

where $\langle \cdots \rangle$ denotes the volume average of $\cdots$. A similar definition [66] was used for a composite material which is subjected to a homogeneous external electric field. In view of the real quantities under consideration, Eq. (7.6) can be rewritten as

$$\langle \mathbf{D}_c \rangle = \varepsilon_e \langle \mathbf{E} \rangle + 4\pi\chi \langle \mathbf{E} \rangle^2 \langle \mathbf{E} \rangle. \tag{7.7}$$

On the other hand, based on thermodynamics the effective dielectric constant ε_E including the incremental part due to the light-striction is defined as

$$\varepsilon_E \equiv \left(\frac{\partial \langle \mathbf{D}_c \rangle}{\partial \langle \mathbf{E} \rangle} \right)_{T,p} = \left(\frac{\partial \langle \mathbf{D}_c \rangle}{\partial \langle \mathbf{E} \rangle} \right)_{T,\rho} + \int f(d) \left(\frac{\partial \langle \mathbf{D}_c \rangle}{\partial \rho(d)} \right)_{T,\langle \mathbf{E} \rangle} \left(\frac{\partial \rho(d)}{\partial \langle \mathbf{E} \rangle} \right)_{T,p} dd, \tag{7.8}$$

where $\rho(d)$ stands for the density of the particles with diameter d, and T temperature. Here $\left(\frac{\partial \langle \mathbf{D}_c \rangle}{\partial \langle \mathbf{E} \rangle} \right)_{T,\rho}$ corresponds to the effective linear dielectric constant, namely ε_e. In Eq. (7.8), $f(d)$ denotes a specific size distribution which exists in real electrorheological nanofluids [179, 180], e.g. the lognormal distribution $f(d) = \frac{1}{\sqrt{2\pi}\sigma d} \exp[-\frac{\ln^2(d/\delta)}{2\sigma^2}]$, where σ is the standard deviation and δ the median diameter.

Accordingly, the incremental dielectric constant due to the light-striction [the last term of Eq. (7.8)] is equivalent to $12\pi\chi \langle \mathbf{E} \rangle^2$. That is,

$$12\pi\chi \langle \mathbf{E} \rangle^2 = \int f(d) \left(\frac{\partial \langle \mathbf{D}_c \rangle}{\partial \rho(d)} \right)_{T,\langle \mathbf{E} \rangle} \left(\frac{\partial \rho(d)}{\partial \langle \mathbf{E} \rangle} \right)_{T,p} dd. \tag{7.9}$$

Let us take one step forward to rewrite Eq. (7.9) as

$$\chi \langle \mathbf{E} \rangle^2 = \frac{1}{12\pi} \int f(d) \langle \mathbf{E} \rangle \left(\frac{\partial \varepsilon_e}{\partial \rho(d)} \right)_{T,\langle \mathbf{E} \rangle} \left(\frac{\partial \rho(d)}{\partial \langle \mathbf{E} \rangle} \right)_{T,p} dd, \tag{7.10}$$

The differential increase of the density inside the light-affected volume $d\rho(d)$ corresponds to the increase in mass equal to $V_c d\rho(d)$. Naturally, this increase in mass is equal to a decrease in mass outside the light-affected volume, which is given by $-\rho(d)d(V - V_c) = -\rho(d)dV$, so that $d\rho(d) = -[\rho(d)/V_c]dV$. Consequently, we may rewrite Eq. (7.10) as

$$\chi \langle \mathbf{E} \rangle^2 = -\frac{1}{12\pi} \int f(d) \langle \mathbf{E} \rangle \frac{\rho(d)}{V_c} \left(\frac{\partial \varepsilon_e}{\partial \rho(d)} \right)_{T,\langle \mathbf{E} \rangle} \left(\frac{\partial V}{\partial \langle \mathbf{E} \rangle} \right)_{T,p} dd. \tag{7.11}$$

Next, we can obtain $\left(\frac{\partial V}{\partial \langle \mathbf{E} \rangle} \right)_{T,p}$ by using the differential of the free energy dF

$$dF = -pdV - SdT + \frac{V_c}{4\pi} \langle \mathbf{E} \rangle d\langle \mathbf{D}_c \rangle, \tag{7.12}$$

where S denotes the entropy. In view of the transformed free enthalpy G

$$G = F + pV - \frac{V_c}{4\pi} \langle \mathbf{E} \rangle \langle \mathbf{D}_c \rangle, \tag{7.13}$$

the differential of G admits the form

$$dG = -SdT + Vdp - \frac{V_c}{4\pi} \langle \mathbf{D}_c \rangle d\langle \mathbf{E} \rangle. \tag{7.14}$$

Based on this equation, we obtain

$$\left(\frac{\partial V}{\partial \langle \mathbf{E} \rangle}\right)_{T,p} = -\frac{V_c \langle \mathbf{E} \rangle}{4\pi} \left(\frac{\partial \varepsilon_e}{\partial p}\right)_{T,\langle \mathbf{E} \rangle}. \tag{7.15}$$

Then, the substitution of Eq. (7.15) to Eq. (7.11) yields

$$\chi \langle \mathbf{E} \rangle^2 = \frac{1}{48\pi^2} \int f(d) \langle \mathbf{E} \rangle^2 \rho(d) \left(\frac{\partial \varepsilon_e}{\partial \rho(d)}\right)_{T,\langle \mathbf{E} \rangle} \left(\frac{\partial \varepsilon_e}{\partial p}\right)_{T,\langle \mathbf{E} \rangle} dd. \tag{7.16}$$

Now let us use

$$\left(\frac{\partial \varepsilon_e}{\partial p}\right)_{T,\langle \mathbf{E} \rangle} = \beta \rho(d) \left(\frac{\partial \varepsilon_e}{\partial \rho(d)}\right)_T, \tag{7.17}$$

where

$$\beta = -\frac{1}{V} \left(\frac{\partial V}{\partial p}\right)_T \tag{7.18}$$

denotes the compressibility in the absence of the external electric field. For deriving Eq. (7.17), we have neglected the terms which depends on $\langle \mathbf{E} \rangle$ because they lead to terms in powers of $\langle \mathbf{E} \rangle$ higher than the second in Eq. (7.16). In the light of the same approximation, the substitution of Eq. (7.17) into Eq. (7.16) leads to

$$\chi \langle \mathbf{E} \rangle^2 = \frac{1}{48\pi^2} \int f(d) \langle \mathbf{E} \rangle^2 \beta \rho(d)^2 \left(\frac{\partial \varepsilon_e}{\partial \rho(d)}\right)_T^2 dd. \tag{7.19}$$

So far, the effective third-order nonlinear susceptibility χ of the electrorheological nanofluid is given by

$$\chi = \frac{\beta}{48\pi^2} \int f(d) \rho(d)^2 \left(\frac{\partial \varepsilon_e}{\partial \rho(d)}\right)_T^2 dd. \tag{7.20}$$

For determining the effective linear dielectric constant ε_e, we can resort to the anisotropic Maxwell-Garnett theory, namely

$$\frac{g_L(\varepsilon_e - \varepsilon_2)}{\varepsilon_2 + g_L(\varepsilon_e - \varepsilon_2)} = \frac{4\pi}{3} \int f(d) \frac{\rho(d)}{m(d)} \alpha(d) dd, \tag{7.21}$$

where $m(d)$ $(\alpha(d))$ denotes the mass (polarizability) of the individual particle with diameter d, and ε_2 the dielectric constant of the carrier liquid. It is known that in the presence of an electric field, the particle chain can be formed in the direction of the field, and thus the structural anisotropy should appear inside this electrorheological nanofluid. Accordingly, in Eq. (7.21) g_L $(g_L \geq 1/3)$ is the local field factor in the longitudinal field case, which was measured by using computer simulations [156, 157], satisfying the sum rule $g_L + 2g_T = 1$ [181]. Here g_T represents the local field factor in the transverse field case. As $g_L = 1/3$, the usual Clausius-Mossotti equation recovers, which is valid for an isotropic system. In fact, the degree of anisotropy of the present system is measured by how g_L is deviated from $1/3$.

Eq. (7.20) is the main result of this section. In detail, the light-striction-induced third-order nonlinear susceptibility χ can be expressed in terms of the size distribution function and density of the particles, the effective linear dielectric constant, etc. In particular, it is apparent to see that at constant pressure χ is proportional to the compressibility of the electrorheological nanofluids of interest. More precisely, χ is of about the same order of magnitude as the compressibility, which can be readily measured in experiments. Interestingly, Eq. (7.20) has bridged the mechanical properties and nonlinear optical properties of the linear electrorheological nanofluids. In other words, the mechanical properties give rise to the nonlinear optical responses (third-order nonlinear susceptibilities) of the linear electrorheological nanofluids.

In what follows, we would like to show the correctness of the present theory by comparing with a different statistical method. First, let us derive the increase of the density $\Delta\rho$ due to light-striction, based on $(\partial V/\partial\langle\mathbf{E}\rangle)_{T,p}$. Let us start from

$$\Delta\rho = \int\int_0^{\langle\mathbf{E}\rangle} f(d)\left(\frac{\partial\rho(d)}{\partial\langle\mathbf{E}\rangle}\right)_{T,p} \mathrm{d}\langle\mathbf{E}\rangle \mathrm{d}d. \tag{7.22}$$

To this end, we obtain

$$\Delta\rho = \frac{1}{8\pi}\int f(d)\langle\mathbf{E}\rangle^2\beta\rho(d)^2\left(\frac{\partial\varepsilon_e}{\partial\rho(d)}\right)_T \mathrm{d}d. \tag{7.23}$$

Again, in the expression for $\Delta\rho$ terms in powers of $\langle\mathbf{E}\rangle$ higher than the second have been neglected. For a monodisperse case, Eq. (7.23) reduces to

$$\Delta\rho = \frac{1}{8\pi}\langle\mathbf{E}\rangle^2\beta\rho^2\left(\frac{\partial\varepsilon_e}{\partial\rho}\right)_T. \tag{7.24}$$

Let us assume there is an ideal gas inside the light-affected volume. In this case, the compressibility is given by

$$\beta = \frac{M}{\rho RT}, \tag{7.25}$$

where M is the molecular weight, and R the molar gas constant. For the ideal gas (monodisperse case), setting $g_L = 1/3$ to the above Clausius-Mossotti equation [Eq. (7.21)] yields

$$\frac{\varepsilon_e - 1}{\varepsilon_e + 2} = \frac{4\pi}{3}\frac{\rho}{m}\alpha. \tag{7.26}$$

In view of $\varepsilon_e - 1 \ll 1$ for ideal gases, we obtain

$$\left(\frac{\partial\varepsilon_e}{\partial\rho}\right)_T = \frac{4\pi}{m}\alpha, \tag{7.27}$$

and hence the desired results for $\Delta\rho$,

$$\Delta\rho = \frac{\langle\mathbf{E}\rangle^2\rho\alpha}{2k_BT}. \tag{7.28}$$

This equation can also be achieved by using a statistical method. According to Boltzmann's distribution law, the number of moles per cm^3 of the gas at a point with field strength $\langle\mathbf{E}\rangle$ is given by

$$N = N'\exp(-\frac{W}{k_BT}), \tag{7.29}$$

where W denotes the average value of the work required to bring a molecule into the field $\langle \mathbf{E} \rangle$, and N' the number of moles per cm^3 of the gas at a point in the absence of field. It is straightforward to obtain

$$\Delta \rho = M(N - N') = \frac{\langle \mathbf{E} \rangle^2 \rho \alpha}{2 k_B T}, \qquad (7.30)$$

which is exactly the same as Eq. (7.28). Again, the terms in higher powers of $\langle \mathbf{E} \rangle$ than the second have been neglected.

To sum up, by using thermodynamics we have presented a theoretical approach to the derivation of the effective third-order nonlinear susceptibility [Eq. (7.20)] of the linear electrorheological nanofluids under the influence of light-striction, which is of about the same order of magnitude as the compressibility of the electrorheological nanofluid at constant pressure. Our approach has been demonstrated in excellent agreement with an alternative statistical method.

In conclusion, the aim of the present section is to exploit light-striction in a linear electrorheological nanofluid (or ferrofluid) in order to generate nonlinear optical responses. Thus, we have shown theoretically that the linear electrorheological nanofluids (or ferrofluids) under the influence of the light-striction effect can serve as a new nonlinear optical material.

Chapter 8

Review of Other Nonlinear Optical Materials

In this Chapter, we shall review briefly the progress in the design of new nonlinear optical materials based on organic and polymeric materials, and inorganic materials. So far, various approaches have been developed to explore new nonlinear optical materials based on them.

1. Based on Organic and Polymeric Materials

Organic molecules and polymers have emerged as a new class of highly promising nonlinear optical materials which has captured the attention of scientists world [182]. Unlike inorganic systems, where nonlinear optical phenomena arise from band structure effects, in organic and polymer systems, nonlinear optical effects originate in the virtual electron excitations occurring on the individual molecular, or polymer chains units. Owing to their large optical nonlinearities and mechanical, chemical, thermal and photo stabilities, organic and polymeric nonlinear optical materials, which have been the subject of very intensive studies, are the leading practical materials for optoelectronic and electronic devices, integrated optical devices, image processing, and optical communications [183, 184, 185, 186, 187, 188, 189, 190].

The known strategy in molecular and crystal engineering of solid nonlinear optical materials is synthesis of organic chromophores with chiral centers [191]. A new polymorphic modification obtained by crystallization from acetonitrile solution was discovered for 2-adamantylamino-5-nitropyridine, a prospective nonlinear optical material [191]. This new phase is characterized by the optimal molecular orientation for the highest effective nonlinear optical responses in the solid state [191].

It is known that nonlinear optical chromophores are very often highly polar in their ground states, therefore it is reasonable to expect that the dipole-dipole interactions between polar molecules would favor antiparallel head-to-tail packing, and hence lower nonlinear characteristics even in the case of acentric crystals. In order to reduce the probability of antiparallel (namely, centrosymmetric) packing, a number of molecules in which neutral bulky groups alternate with the hyperpolarizable moety have been prepared [192, 193]. Following this idea a large series of 2-amino-5-nitropyridine derivatives with different

143

bulky substituents at the amino-nitrogen atom have been obtained to demonstrate the workability of this approach in nonlinear optical crystal engineering [192, 193, 194]. 2-cyclooctylamino-5-nitropyridine was first described as a new prospective nonlinear optical material in 1987 [192]. It was found [195] that this compound exhibits an anomalous solidification upon cooling from the melt which allows for the formation of a poling induced non-centrosymmetric glassy state exhibiting pronounced second harmonic generation and large nonlinear optical susceptibilities. This compound is of importance in fabrication of optically nonlinear glassy thin films for guided wave optics.

Many other derivatives of 2-amino-5-nitropyridine and 2-amino-5-nitropyridinium slats have also been found to possess second harmonic generation properties in the solid state, forming "optimal" or close to optimal herringbone motifs in their crystals [193, 194, 196]. Therefore, this fragment may be considered as a good synthon for nonlinear optical crystals engineering. Among these derivatives are numerous salts of the 2-amino-5-nitropyridinium cation with different inorganic anions ($H_2PO_4^-$, $H_2AsO_4^-$, $CHCl_2COO^-$, Cl^-, Br^-, $HO_3PCH_2COO^-$) [194, 196] and 2-adamantylamino-5-nitropyridine. The latter was found to be a very active solid nonlinear optical material, whose relative second harmonic generation intensity is about 300 times of that of urea [197].

Using pore orientations to produce optical materials was attempted by doping mesoporous materials with semiconducting polymers [198]. A new approach to second-order nonlinear optical materials was reported [199], in which chirality and supramolecular organization play key roles. The chiral supramolecular organization makes the second-order nonlinear optical susceptibility about 30 times larger for the nonracemic material than for the racemic material with the same chamical structure.

A new linear epoxy polymer containing 4-amino-4-nitrotolane chromophores attached to the chain backbone was synthesized in an attempt to enhance poling-induced nonlinear optical susceptibilities [200]. Suslick *et al.* [201] have synthesized a series of "puch-pull" porphyrins containing both donor and acceptor substituents and showed them to serve as nonlinear optical materials. One used very short and linear C(sp)-H$\cdots$N hydrogen bonds to form head-to-tail straight tapes and their assembly, in an attempt to obtain nonlinear optical polar crystals [202].

Crystal-engineering strategies were developed toward the synthesis of noncentrosymmetric infinite coordination networks for use as second-order nonlinear optical materials [203].

New organometallic donor-acceptor molecules were synthesized for second-order nonlinear optics [204], and they contain ferrocene as the donor group, cyano or tricyano-derivatized furan as the acceptor, and thiophene or 3,3'-bipyridine derivatives as pi-bridge.

Organic-organic cocrystals were prepared from 2-amino-5-nitropyridine and achiral benzenesulfonic acids, which were designed for second-order nonlinear optical materials [205]. Thermally stable heterocyclic imines were proposed as new potential nonlinear optical materials [206].

High quality bulk single crystals of semiorganic nonlinear optical materials, L-arginine tetrafluoroborate and L-histidine tetrafluoroborate, was grown from solution by temperature lowering methods [207].

Azaphosphane-based chromophores were developed with nonlinear optical activities [208]. Here azaphosphanes serve as first examples of efficient electron donors in the chemical architecture of new nonlinear optical materials.

One designed nonlinear optical materials based on interpenetrated diamondoid coordination networks [209] and 2D coordination networks [210].

2. Based on Inorganic Materials

The ternary BaO-TiO_2-B_2O_3 glasses containing a large amount of TiO2 (20-40 mol%) were prepared, and their optical basicities, the formation, structural features and second-order optical nonlinearities of $BaTi(BO_3)_2$ and $Ba_3Ti_3O_6(BO_3)_2$ crystals were examined to develop new nonlinear optical materials [211]. Another development [212] of new materials with large nonlinear susceptibilities is the exploration of quasi-one-dimensional systems, or quantum wires ł the quantum confinement of electron-hole motion in one-dimensional space can enhance third-order nonlinear susceptibilities $\chi^{(3)}$, e.g., one-dimensional MottCHubbard insulators [212].

An important approach [191] in the discovery of new nonlinear optical crystals is a systematic search for nonlinear-optical-active polymorphs using different methods of single crystal growth (crystallization from different solvents and/or under different conditions, from melts, vapor phase, etc.). The existence of polymorphic forms provides an opportunity to obtain new nonlinear-optical-active forms. In particular, two nolinear-optical-active forms were found for the organic chromophore 8-(4'-acetylphenyl)-1,4-dioxa-8-azaspiro[4.5]decane using the melt growth and plate sublimation techniques [213]. Antipin *et al.* [191] searched for new possible polymorphic phases of these compounds by using different techniques for crystal growth.

For obtaining new nonlinear optical materials, one applied ion implantation for synthesis of copper nanoparticles in a zinc oxide matrix [214]. The new nonlinear optical material comprising metal nanoparticles in a ZnO matrix exhibits the phenomenon of self-defocusing and possesses a high nonlinear absorption coefficient.

A new nonlinear optical crystal CsLiBO [215] was described that can be grown from either stoichiometric melt or from solution. A large, high quality single crystal with dimensions of $14 \times 11 \times 11$ cm^3 was obtained by the top-seeded Kyropoulos method. Fourth harmonic and fifth harmonic generations of the $1.064\,\mu m$ Nd:YAG laser radiation with type-I phase matching were realized in the CsLiBO crystal.

The new nonlinear optical crystal $KTiOAsO_4$ [216] was characterized and evaluated for use in optical parametric oscillator devices capable of generating tunable radiation in the $3C5\,\mu m$ region.

A different strategy for the design of nonlinear optical materials is to use solid-state proton transfer for achieving nonlinearity enhancement [217].

Bis thiourea zinc chloride was synthesized as a new semiorganic nonlinear optical material [218]. Single crystals of the compound had been grown by slow evaporation of saturated aqueous solution at room temperature.

Finally, mesostructured and mesoporous materials are also emerging as a new class of nonlinear optical materials [219]. Another mechanism for generating optical mesostructured materials is to utilize the organized pores as hosts for growing or depositing inorganic materials. It is well-known that as semiconducting materials decrease in size, their optical properties drastically change as quantum confinement occurs [220]. Specifically, the idea is to use the mesopores, $4 - 50\,$nm, to generate ordered arrays of semiconductor nanocrystals [219] (or luminescent organometallic complexes [221]).

Chapter 9

Summary and Outlook

Composite effects are always expected to open a fascinating field of new phenomena in nonlinear optics. We have presented an original, and first-handed review of the state-of-the-art development on the design of new nonlinear optical materials, with an emphasis on understanding the physical processes of the composite effects on the enhancement of optical nonlinearity of the materials. Besides third-order nonlinear optical susceptibilities, we have also considered a number of optical processes, e.g., four-wave mixing, second or third harmonic generation, etc. We have investigated composite effects on such nonlinear optical processes, which are interesting and useful in the sense that they are accessible to experimentalists working in the field.

Since strong fluctuations of local fields may result in large optical nonlinearities in nanostructured composites, composite materials can have larger nonlinear susceptibilities at zero and finite frequencies than those of ordinary bulk materials or constituent materials from which the composite is constructed. The formation of composite materials thus constitutes a means for engineering new materials with desired enhanced nonlinear optical response [20]. Basically, the response of a nonlinear composite can be tuned by controlling the volume fraction and morphology of constitutes. The latter can be purposefully adjusted by using external electric/magnetic field, by adding gradation profile to the system of interest, and so on. In this direction, several classes of nonlinear optical materials were designed, namely,

1. Colloidal nanocrystals with inhomogeneous metallodielectric particles or a graded-index host

Such materials can have both an enhancement and a red shift of optical nonlinearity, due to the gradation inside the metallic core or host as well as the lattice effects arising from the periodic structure.

2. Metallic films with inhomogeneous microstructures adjusted by ion doping or temperature gradient

It has been found that the presence of gradation (or multilayer) in metallic films yields a broad resonant plasmon band in the optical region, resulting in a large nonlinearity enhancement and hence an optimal figure of merit.

3. Composites with compositional gradation or graded particles

We found enhanced nonlinear optical responses can be achieved due to the presence of gradation inside compositionally graded metal-dielectric films and/or dielectrically-graded particles with/without dielectric anisotropy. A spectral representation was developed to understand the enhanced responses.

4. Magneto-controlled ferrofluid-based nonlinear optical materials

By including a metallic nonlinear shell in field-responsive ferrofluids, one can tune the enhanced nonlinear optical properties by applying an external magnetic field. Such a proposed magneto-controlled nonlinear optical material can serve as optical materials which have anisotropic nonlinear optical properties and a giant enhancement of nonlinearity, as well as an attractive figure of merit.

5. Electrorheological nanofluids or ferrofluids partially subjected to a Gaussian laser beam (light-striction)

We exploited light-striction in linear electrorheological nanofluids or ferrofluids in order to generate nonlinear optical responses, and showed theoretically that the linear electrorheological nanofluids and ferrofluids under the influence of the electrostriction effect can serve as a new nonlinear optical material.

Finally, a survey on the design of nonlinear optical materials by other materials was done. To sum up, the main features of the proposed nonlinear materials are that:

(1) They can offer higher effective nonlinear optical responses and hence desired faster response times, due to the many-body (local-field) effect and the long-range lattice effect in the particle chains/columns or clusters in the system subjected to external fields;

(2) The nonlinear optical responses or the response times can be real-time-adjusted by choosing appropriate external electric or magnetic fields, due to the change of the microstructure in the suspensions;

(3) Cost can be saved much since the suspension-based nonlinear optical materials have fluidity and are difficult to be abraded.

As future agenda, it is possible to design additional optical materials with nonlinearity enhancement by using similar concepts, e.g., using graded shells in field-responsive complex fluids, etc. Computer simulations might also be used to design new nonlinear optical materials. For example, we might do computer simulations to achieve various microstructure of a wide range of complex fluids or colloidal nanocrystals under different conditions, and hence obtain desired effective nonlinear optical responses that correspond to different composite effects arising from local-field effects in the microstructure. Following the approaches mentioned in the review, new types of nonlinear dielectric materials may also be designed [112, 222] for use in electronic and microwave components, sensor windows, and so on.

In: New Nonlinear Optical Materials: Theoretical Research   ISBN 1-60021-402-9
Editors: J.P. Huang and K.W. Yu, pp. 149–151   © 2006 Nova Science Publishers, Inc.

Appendix A

An Illustration of the Bergman-Milton Spectral Representation

The essence of the Bergman-Milton spectral representation is to define the following transformations. If we denote a material parameter

$$s = \left(1 - \frac{\varepsilon_1}{\varepsilon_2}\right)^{-1},$$

(A.1)

then the reduced effective dielectric constant

$$w(s) = 1 - \frac{\varepsilon_e}{\varepsilon_2},$$

(A.2)

can be written as

$$w(s) = \sum_n \frac{F_n}{s - s_n},$$

(A.3)

where n is a positive integer, i.e., $n = 1, 2, ...,$ and F_n and s_n, are the n-th microstructure parameters of the composite materials [26]. In Eq. (A.3), $0 \le s_n < 1$ is a real number, while F_n satisfies a sum rule [26]

$$\sum_n F_n = p.$$

(A.4)

In what follows, we illustrate the spectral representation by the capacitance of simple geometry [28]. In particular, a parallel-plate capacitor is considered as an example. We will discuss two cases, namely, the series combination (Fig. A.1) and the parallel combination (Fig. A.2).

In the first case (see Fig. A.1), if one inserts a dielectric slab of dielectric constant ε_1 and thickness h_1, as well as a dielectric of ε_2 and thickness h_2 (both of the same area A), into a parallel-plate capacitor of total thickness $h = h_1 + h_2$, the overall capacitance C is given by

$$C^{-1} = C_1^{-1} + C_2^{-1},$$

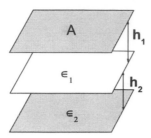

Figure A.1. Schematic graph showing a parallel-plate capacitor of total thickness $h = h_1 + h_2$ that contains a dielectric slab of dielectric constant ε_1 and thickness h_1, as well as a dielectric of ε_2 and thickness h_2 (both of the same area A).

where $C_1 = \varepsilon_1 A/h_1$ and $C_2 = \varepsilon_2 A/h_2$. On the other hand, we may define the equivalent capacitance as $C = \varepsilon_e A/h$, where ε_e is the effective dielectric constant. That is, we see the composite dielectric as a homogeneous dielectric of dielectric constant ε_e.

Let $\varepsilon_1 = \varepsilon_2(1 - 1/s)$, we can express C in the spectral representation,

$$C = \frac{A\varepsilon_2}{h} - \frac{A\varepsilon_2 h_1/h^2}{s - h_2/h}.$$

In accord with the spectral representation, one may introduce $w(s) = 1 - \varepsilon_e/\varepsilon_2$, which is in fact the same as $w(s) = 1 - C/C_0$, where C_0 is the capacitance when the plates are all filled with a dielectric material of ε_2, namely $C_0 = \varepsilon_2 A/h$. Thus we obtain

$$w(s) = \frac{h_1/h}{s - h_2/h}.$$

from which we find that the material parameter is separated from the geometric parameter. The comparison of $w(s)$ with Eq.(A.3) yields

$$F_1 = h_1/h, \quad s_1 = h_2/h.$$

It is worth noting that F_1 obtained herein is just equal to the volume fraction of the dielectric of ε_1, and that s_1 satisfies $0 \le s_1 < 1$, as required by the spectral representation theory.

Next, we consider the parallel combination (see Fig. A.2). If one inserts a material with dielectric constant ε_1 and area w_1 as well as a dielectric with ε_2 and w_2 (both of the same thickness h), into a parallel-plate capacitor of total area $A = w_1 + w_2$, the overall capacitance C is given by

$$C = C_1 + C_2,$$

where $C_1 = \varepsilon_1 w_1/h$ and $C_2 = \varepsilon_2 w_2/h$. Similarly, after introducing the effective dielectric constant ε_e, we may define the overall capacitance as $C = \varepsilon_e A/h$.

Again, in the spectral representation, let $s = (1 - \varepsilon_1/\varepsilon_2)^{-1}$, then

$$C = \frac{\varepsilon_2 A}{h} - \frac{\varepsilon_2 w_1}{hs}.$$

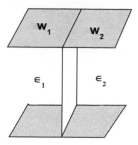

Figure A.2. Schematic graph showing a parallel-plate capacitor of total area $w = w_1 + w_2$ that contains a dielectric slab of dielectric constant ε_1 and area w_1, as well as a dielectric of ε_2 and area w_2 (both of the same thickness h).

Writing $w(s) = 1 - C/C_0$, we obtain

$$w(s) = \frac{w_1/A}{s}.$$

From this equation, the material parameter is also found to be separated from the geometric parameter. It is clear that $F_1 = w_1/A$, i.e., the volume fraction of the dielectric of ε_1, and $s_1 = 0$.

In: New Nonlinear Optical Materials: Theoretical Research   ISBN 1-60021-402-9
Editors: J.P. Huang and K.W. Yu, pp. 153–154   © 2006 Nova Science Publishers, Inc.

Appendix B

Depolarization Factors of Particles

1. General ellipsoidal particles

For the general ellipsoid with the three principal axes a, b, and c. Here $a > b > c$, the depolarization factors L_a, L_b and L_c depend on the axis ratios $\beta = b/a$ and $\delta = c/a$ and are given by [223, 224]

$$L_a = \frac{\beta\delta}{\sqrt{1-\delta^2}(1-\beta^2)}(\rho_1(k,\phi) - \rho_2(k,\phi)) \tag{B.1}$$

$$L_b = -L_a + \frac{\beta\delta}{\sqrt{1-\delta^2}(\beta^2-\delta^2)}\rho_2(k,\phi) - \frac{\delta^2}{\beta^2-\delta^2} \tag{B.2}$$

$$L_c = -\frac{\beta\delta}{\sqrt{1-\delta^2}(\beta^2-\delta^2)}\rho_2(k,\phi) + \frac{\beta^2}{\beta^2-\delta^2} \tag{B.3}$$

ρ_1 and ρ_2 are the elliptical integrals which are functions of k and ϕ

$$\rho_1(k,\phi) = \int_0^\phi \frac{1}{\sqrt{1-k^2\sin^2\Phi}}d\Phi, \quad \rho_2(k,\phi) = \int_0^\phi \sqrt{1-k^2\sin^2\Phi}\,d\Phi,$$

where

$$k = \sqrt{\frac{1-\beta^2}{1-\delta^2}}, \quad \phi = \arccos(\delta).$$

2. Spheroidal particles

For the spheroid with two equal axes a and b, and symmetry axis c, it is possible to obtain explicit expressions for the depolarization factors [225], $L_a(=L_b)$ and L_c.

For the prolate spheroid case $a < c$, one obtains

$$L_c = \frac{1}{1-q^2} + \frac{q}{(q^2-1)^{3/2}}\ln\left(q+\sqrt{q^2-1}\right), \tag{B.4}$$

with aspect ratio $q = c/a > 1$.

For the oblate spheroid case $a > c$, one obtains

$$L_c = \frac{1}{1-q^2} + \frac{q}{(1-q^2)^{3/2}}\arccos q, \tag{B.5}$$

with aspect ratio $q = c/a < 1$.

In Eqs.(B.4) and (B.5), q denotes the eccentricity of the spheroid. Note for general ellipsoids including spheroids, the depolarization factors along the three principal axes must satisfy the sum rule

$$L_a + L_b + L_c = 1. \tag{B.6}$$

Thus, for the spheroid, the depolarization factor along $a-$ (or $b-$) axis is

$$L_a \equiv L_b = \frac{1 - L_c}{2}. \tag{B.7}$$

3. Spherical particles

In fact, spheres are a special case of spheroids with $a = b = c$. According to Eqs. B.6 and B.7, we have the depolarization factors

$$L_a = L_b = L_c = \frac{1}{3}. \tag{B.8}$$

4. Spheroidal cylinders

Let us assume the symmetry axis of a spheroidal cylinder to be in the $z-$axis. Thus, we have the depolarization factors

$$L_x = 1 - L_y, \; L_z = 0, \tag{B.9}$$

where L_x and L_y are the depolarization factors in the $xy-$plane.

5. Circular cylinders

Circular cylinders are a special case of spheroidal cylinders with $L_x = L_y$, thus according to Eq. (B.9) we have

$$L_x = L_y = 0.5, \; L_z = 0. \tag{B.10}$$

6. Disks

For a disk, if we assume the symmetry axis of the disk to be in the $z-$axis, we obtain the depolarizatin factors

$$L_x = L_y = 0, \; L_z = 1 \tag{B.11}$$

where L_x and L_y are again the depolarization factors in the $xy-$plane.

In: New Nonlinear Optical Materials: Theoretical Research   ISBN 1-60021-402-9
Editors: J.P. Huang and K.W. Yu, pp. 155–156   © 2006 Nova Science Publishers, Inc.

Appendix C

Differential Effective Dipole Theory

If we add to a homogeneous spherical core of dielectric constant ε_1 a spherical shell of ε, to make a coated sphere [226] of overall radius a_1. The dipole factor of the coated sphere b_1 is [227],

$$b_1 = \frac{(\varepsilon - \varepsilon_2) + (\varepsilon_2 + 2\varepsilon)\rho_1 \alpha_1}{(\varepsilon + 2\varepsilon_2) + 2(\varepsilon - \varepsilon_2)\rho_1 \alpha_1}, \tag{C.1}$$

where ρ_1 is given by

$$\rho_1 = \frac{\varepsilon_1 - \varepsilon}{\varepsilon_1 + 2\varepsilon}, \tag{C.2}$$

and

$$\alpha_1 = (a/a_1)^3. \tag{C.3}$$

The consideration can be extended to more shells of different dielectric constants, at the expense of more complicated expressions [227, 228, 43]. It is easy to check that b_1 reduces to b_0 when $\varepsilon = \varepsilon_1$. Thus, the dipole factor remains unchanged if one adds a spherical shell of the same dielectric constant.

Now we develop the differential effective dipole theory (DEDT) for spherical particles of graded materials. To establish the DEDT, we mimic the graded profile by a multi-shell construction, i.e., we build up the dielectric profile gradually by adding shells. We start with an infinitesimal spherical core of dielectric constant $\varepsilon(0)$ and keep on adding spherical shells of dielectric constant given by $\varepsilon(r)$ at radius r, until $r = a$ is reached. At radius r, we have an inhomogeneous sphere whose dipole factor is given by $b(r)$. We further replace the inhomogeneous sphere by a homogeneous sphere of the same dipole factor and the graded profile is replaced by an equivalent dielectric constant $\bar{\varepsilon}(r)$. Thus,

$$b(r) = \frac{\bar{\varepsilon}(r) - \varepsilon_2}{\bar{\varepsilon}(r) + 2\varepsilon_2}. \tag{C.4}$$

Next, we add to the sphere a spherical shell of infinitesimal thickness dr, of dielectric constant $\varepsilon(r)$. The dipole factor will change according to Eq. (C.1). Of course, the equivalent dielectric constant $\bar{\varepsilon}(r)$, being related to $b(r)$, should also change by the same token. Let us write $b_1 = b + db$, and take the limit $dr \to 0$, we obtain a differential equation [65]:

$$\frac{db(r)}{dr} = -\frac{1}{3r\varepsilon_2\varepsilon(r)}[(1 + 2b(r))\varepsilon_2 - (1 - b(r))\varepsilon(r)][(1 + 2b(r))\varepsilon_2 + 2(1 - b(r))\varepsilon(r)]. \tag{C.5}$$

Thus the dipole factor of a graded spherical particle can be calculated by solving the above differential equation with a given graded profile $\varepsilon(r)$. The nonlinear first-order differential equation can be integrated, at least numerically, if we are given the graded profile $\varepsilon(r)$ and the initial condition $b(r = 0)$. The substitution of Eq. C.4 into Eq. C.5 yields the equivalent dielectric constant $\bar{\varepsilon}(r = a)$ for the graded spherical particle, which satisfies the differential equation

$$\frac{d\bar{\varepsilon}(r)}{dr} = \frac{[\varepsilon(r) - \bar{\varepsilon}(r)][\bar{\varepsilon}(r) + 2\varepsilon(r)]}{r\varepsilon(r)}. \tag{C.6}$$

Finally, we should remark that Eq. (C.5) [and hence Eq. (C.6)] for the DEDT is not only valid but also exact for arbitrary gradation profiles, as it has been demonstrated in perfect agreement with a first-principles approach [229].

In: New Nonlinear Optical Materials: Theoretical Research   ISBN 1-60021-402-9
Editors: J.P. Huang and K.W. Yu, pp. 157–158   © 2006 Nova Science Publishers, Inc.

Appendix D

Anisotropic Differential Effective Dipole Theory

Let us consider a graded spherical particle with radius a. We adopt the spherical coordinates for convenience. The graded spherical particle has a tangential dielectric constant in the plane orthogonal to the radial vector of the sphere ($\varepsilon_{\theta\theta}(r) = \varepsilon_{\phi\phi}(r)$), and a radial dielectric constant $\varepsilon_{rr}(r)$. Both $\varepsilon_{\theta\theta}(r)$ and $\varepsilon_{rr}(r)$ will be prescribed by radial functions. In view of the symmetry, the anisotropic dielectric constant of the graded sphere can be expressed as tensor $\bar{\bar{\varepsilon}}_c(r)$ [135, 133], namely,

$$\bar{\bar{\varepsilon}}_c(r) = \begin{pmatrix} \varepsilon_{rr}(r) & 0 & 0 \\ 0 & \varepsilon_{\theta\theta}(r) & 0 \\ 0 & 0 & \varepsilon_{\phi\phi}(r) \end{pmatrix}. \tag{D.1}$$

It is worth noting that the above form is in spherical coordinates, rather than in Cartesian coordinates.

Now we present an anisotropic differential effective dipole theory (ADEDT), which is a numerical method for the analysis of the dielectric property of anisotropic graded particles with *arbitrary* gradation profiles. We may regard the gradation profile as a multi-shell construction. In details, we establish the dielectric profile gradually by adding shells. Let us start with an infinitesimal isotropic spherical core with dielectric constant $\varepsilon(0)$, and keep on adding shells with both tangential and normal dielectric-constant profiles $\varepsilon_{\theta\theta}(r)$ and $\varepsilon_{rr}(r)$ at radius r, until $r = a$ is reached. At radius r, we have an inhomogeneous particle, and further regard such an inhomogeneous particle as an effective *homogeneous* one, which has the dipole factor

$$b(r) = \frac{\bar{\varepsilon}(r) - \varepsilon_m}{\bar{\varepsilon}(r) + 2\varepsilon_m} \tag{D.2}$$

where $\bar{\varepsilon}$ is the equivalent dielectric constant of the effective *homogeneous* particle with radius r. It is worth noting that even though a tensorial form was used in Eq. (D.1), the resulting $\bar{\varepsilon}$ should still be scalar. Then, we add to the particle a shell with infinitesimal thickness $\triangle r$, with dielectric constants $\varepsilon_{\theta\theta}(r)$ and $\varepsilon_{rr}(r)$. The dipole factor should change according to the dipole factor of one shell anisotropic composite inclusion [135], that is,

$$b(r+\triangle r) = \frac{(\delta_1(r)\varepsilon_{rr}(r) + \bar{\varepsilon}(r))(\delta(r)\varepsilon_{rr}(r) - \varepsilon_m) + \Pi(\delta_1(r)\varepsilon_{rr}(r) + \varepsilon_m)\rho_r}{(\delta_1(r)\varepsilon_{rr}(r) + \bar{\varepsilon}(r))(\delta(r)\varepsilon_{rr}(r) + 2\varepsilon_m) + \Pi(\delta_1(r)\varepsilon_{rr}(r) - 2\varepsilon_m)\rho_r}, \tag{D.3}$$

with $\Pi = \bar{\varepsilon}(r) - \delta(r)\varepsilon_{rr}(r)$ and $\rho_r = [(r + \triangle r)/r]^{2\delta(r)+1}$, where $\delta(r) = [-1 + (1 + 8\varepsilon_{\theta\theta}(r)/\varepsilon_{rr}(r))^{1/2}]/2$ and $\delta_1(r) = 1 + \delta(r)$. Let us write further $\triangle b(r) = b(r + \triangle r) - b(r)$, and take the limit $\triangle r \to 0$, then the desired correction $\triangle b(r)$ is infinitesimal accordingly. Thus, one can obtain a differential equation as [46]

$$
\begin{aligned}
\frac{db(r)}{dr} &= \frac{1}{3r\varepsilon_{rr}(r)\varepsilon_m} \Big[(1 - b(r))^2 \varepsilon_{rr}(r)^2 \delta(r)(1 + \delta(r)) \\
&\quad - (1 + b(r) - 2b(r)^2)\varepsilon_{rr}(r)\varepsilon_m - (1 + 2b(r))^2 \varepsilon_m^2 \Big],
\end{aligned}
\tag{D.4}
$$

where $0 < r \leq a$. Therefore, the dipole factor of an anisotropic graded spherical particle can be calculated by solving the first-order differential equation [Eq. (D.4)]. This differential equation can be integrated, at least numerically, as long as the gradation profiles ($\varepsilon_{\theta\theta}(r)$ and $\varepsilon_{rr}(r)$) and the initial condition ($b(0)$) are given. After substituting Eq. (D.2) into Eq. (D.4), we obtain the differential equation for the equivalent dielectric constant

$$
\frac{d\bar{\varepsilon}(r)}{dr} = \frac{2\varepsilon_{rr}(r)\varepsilon_{\theta\theta}(r) - \varepsilon_{rr}(r)\bar{\varepsilon}(r) - \bar{\varepsilon}(r)^2}{r\varepsilon_{rr}(r)}.
\tag{D.5}
$$

This is actually the Tartar formula, derived for assemblages of spheres with varying radial and tangential conductivities [43].

The ADEDT [Eq. (D.4) or Eq. (D.5)] is an exact theory for anisotropic graded spherical particles with arbitrary gradation profiles as it was demonstrated in perfect agreement with a first-principles approach [46].

In: New Nonlinear Optical Materials: Theoretical Research   ISBN 1-60021-402-9
Editors: J.P. Huang and K.W. Yu, pp. 159–160   © 2006 Nova Science Publishers, Inc.

Appendix E

Differential Effective Multipole Moment Theory

Now we develop the differential effective multipole moment theory (DEMMT) for a graded spherical particle. To establish the DEMMT, we mimic the graded profile by a multi-shell construction [64], i.e., we build up the dielectric-constant profile gradually by adding shells. We start with an infinitesimal spherical core of dielectric constant $\varepsilon_i(0^+)$ and keep on adding spherical shells of dielectric constant given by $\varepsilon_i(r)$ at radius r, until $r = a$ is reached. At radius r, we have an inhomogeneous sphere with certain multipole moment. We further replace the inhomogeneous sphere by a homogeneous sphere of the same multipole moment and the graded profile is replaced by an equivalent dielectric constant $\bar{\varepsilon}_i(r)$. Thus, the multipole factor $H_l(r)$

$$H_l(r) = \frac{l(\bar{\varepsilon}_i(r) - \varepsilon_m)}{l(\bar{\varepsilon}_i(r) + \varepsilon_m) + \varepsilon_m}. \tag{E.1}$$

Next, we add to the sphere a spherical shell of infinitesimal thickness dr, of dielectric constant $\varepsilon_i(r)$. The resulting multipole factor H_l' will change according to [230]

$$H_l' = \frac{(1 - \varepsilon_m/\varepsilon_i(r))\varepsilon_i(r)\Theta + \rho\varepsilon_i(r)[l'(\bar{\varepsilon}_i(r) - \varepsilon_i(r)) + l\varepsilon_m(\bar{\varepsilon}_i(r)/\varepsilon_i(r) - 1)]}{[1 + l'\varepsilon_m/(l\varepsilon_i(r))]\varepsilon_i(r)\Theta + l'\rho\varepsilon_i(r)(\varepsilon_m - \varepsilon_i(r) + \bar{\varepsilon}_i(r) - \varepsilon_m\bar{\varepsilon}_i(r)/\varepsilon_i(r))} \tag{E.2}$$

with $\Theta = (l'\varepsilon_i(r) + l\bar{\varepsilon}_i(r))$, $l' = l + 1$ and $\rho = [r/(r + dr)]^{2l+1}$.

Of course, the equivalent dielectric constant $\bar{\varepsilon}_i(r)$, being related to $H_l(r)$, should also change by the same token. Let us write $H_l' = H_l + dH_l$, and take the limit $dr \to 0$, we obtain a differential equation [231]:

$$
\begin{aligned}
\frac{dH_l(r)}{dr} &= -\frac{1}{(2l+1)r\varepsilon_m\varepsilon_i(r)}[(H_l(r) + l + H_l(r)l)\varepsilon_m + (H_l(r) - 1)l\varepsilon_i(r)] \\
&\times [(H_l(r) + l + H_l(r)l)\varepsilon_m - (H_l(r) - 1)(l+1)\varepsilon_i(r)].
\end{aligned} \tag{E.3}
$$

Thus the multipole factor of a graded spherical particle can be calculated by solving the above differential equation with a given graded profile $\varepsilon_i(r)$. Solving $H_l(r)$ from Eq.(E.3) gives numerical results for the multipole factor. The nonlinear first-order differential equation can be integrated if we are given the graded profile $\varepsilon_i(r)$ and the initial condition $H_l(r = 0)$. In general, it works for arbitrary graded profiles $\varepsilon_i(r)$.

The substitution of the relation Eq.(E.1) into Eq. (E.3) yields the differential equation for the equivalent dielectric constant $\bar{\varepsilon}_i(r = a)$, which satisfies

$$\frac{d\bar{\varepsilon}_i(r)}{dr} = \frac{[\varepsilon_i(r) - \bar{\varepsilon}_i(r)][(l+1)\varepsilon_i(r) + l\bar{\varepsilon}_i(r)]}{r\varepsilon_i(r)}. \tag{E.4}$$

We should remark again that the DEMMT [Eq. (E.3) or Eq. (E.4)] is indeed exact for arbitrary dielectric gradation profiles, as compared to a first-principles approach in Ref. [229].

In: New Nonlinear Optical Materials: Theoretical Research   ISBN 1-60021-402-9
Editors: J.P. Huang and K.W. Yu, pp. 161–162   © 2006 Nova Science Publishers, Inc.

Appendix F

Anisotropic Differential Effective Multipole Moment Theory

Let us consider a graded spherical particle with radius a. We adopt the spherical coordinates for convenience. The graded spherical particle has a tangential dielectric constant in the plane orthogonal to the radial vector of the sphere $\varepsilon_i^\perp(r)$ and a radial dielectric constant $\varepsilon_i^\parallel(r)$. Both $\varepsilon_i^\perp(r)$ and $\varepsilon_i^\parallel(r)$ will be prescribed by radial functions, $0 < r \le a$. In view of the symmetry, the anisotropic dielectric constant of the graded sphere can be expressed as a tensor form $\bar{\varepsilon}_i(r)$, namely,

$$\bar{\varepsilon}_i(r) = \varepsilon_i^\parallel(r)\hat{r}\hat{r} + \varepsilon_i^\perp(r)\hat{\theta}\hat{\theta} + \varepsilon_i^\perp(r)\hat{\phi}\hat{\phi}. \tag{F.1}$$

Next, we present an anisotropic differential effective multipole moment theory (ADEMMT), which, similar to DEMMT, is a numerical method for the analysis of the electric property of anisotropic graded particles with arbitrary gradation profiles. Similarly, we may regard the gradation profile as a multishell construction. In detail, we establish the electric profile gradually by adding shells. Let us start with an infinitesimal isotropic spherical core with dielectric constant $\varepsilon_i(0^+)$, and keep on adding shells with both tangential and normal dielectric profiles $\varepsilon_i^\perp(r)$ and $\varepsilon_i^\parallel(r)$ at radius r, until $r = a$ is reached. At radius r, we have an inhomogeneous particle, and further regard such an inhomogeneous particle as an effective homogeneous one with an equivalent dielectric constant $\bar{\varepsilon}_i(r)$, which has the multipole factor

$$H_l(r) = \frac{l(\bar{\varepsilon}_i(r) - \varepsilon_m)}{l(\bar{\varepsilon}_i(r) + \varepsilon_m) + \varepsilon_m}. \tag{F.2}$$

Then, we add to the particle a shell with infinitesimal thickness dr, with dielectric constants $\varepsilon_i^\perp(r)$ and $\varepsilon_i^\parallel(r)$. Its multipole factor H_l' should change according to the multipole factor of a single-coated particle [232]

$$H_l' = \frac{(\varepsilon_i^\parallel(r)u_- - \bar{\varepsilon}_i(r)l)(\varepsilon_i^\parallel(r)u_+ - \varepsilon_m l) - \rho_l(\varepsilon_i^\parallel(r)u_+ - \bar{\varepsilon}_i(r)l)(\varepsilon_i^\parallel(r)u_- - \varepsilon_m l)}{(\bar{\varepsilon}_i(r)l - \varepsilon_i^\parallel(r)u_+)(\varepsilon_i^\parallel(r)u_- + \varepsilon_m l')\rho_l - (\bar{\varepsilon}_i(r)l - \varepsilon_i^\parallel(r)u_-)(\varepsilon_i^\parallel(r)u_+ + \varepsilon_m l')}, \tag{F.3}$$

with $u_\pm = [-1 \pm \sqrt{1 + 4l(1+l)\varepsilon_i^\perp(r)/\varepsilon_i^\parallel(r)}]/2$, $\rho_l = [r/(r+dr)]^{u_+ - u_-}$, and $l' = 1 + l$. Let us write further $H_l' = H_l + dH_l$, and take the limit $dr \to 0$, we obtain a differential equation [231]

$$\frac{dH_l(r)}{dr} = -\frac{1}{(1+2l)r\varepsilon_m\varepsilon_i{}^{\parallel}(r)}[l\varepsilon_m - u_-\varepsilon_i{}^{\parallel}(r) + H_l(r)(l'\varepsilon_m + u_-\varepsilon_i{}^{\parallel}(r))]$$
$$[l\varepsilon_m - u_+\varepsilon_i{}^{\parallel}(r) + H_l(r)(l'\varepsilon_m + u_+\varepsilon_i{}^{\parallel}(r))]. \qquad (\text{F.4})$$

Thus the multipole factor of an anisotropic graded spherical particle $H_l(r = a)$ can be calculated by solving the nonlinear first-order differential equation [Eq. (F.4)] which can be integrated, at least numerically if we are given the graded profiles $\varepsilon_i^{\perp}(r)$ and $\varepsilon_i{}^{\parallel}(r)$ and the initial condition $H_l(r = 0)$. The substitution of Eq. (F.2) into Eq. (F.4) yields the differential equation for the equivalent dielectric constant

$$\frac{d\bar{\varepsilon}_i(r)}{dr} = \frac{(1+l)\varepsilon_i{}^{\parallel}(r)\varepsilon_i^{\perp}(r) - \varepsilon_i{}^{\parallel}(r)\bar{\varepsilon}_i(r) - l\bar{\varepsilon}_i(r)^2}{r\varepsilon_i{}^{\parallel}(r)}. \qquad (\text{F.5})$$

Eqs. (F.4) and (F.5) can respectively reduce to Eqs. (E.3) and (E.4), as long as there is $\varepsilon_i^{\perp}(r) = \varepsilon_i{}^{\parallel}(r) = \varepsilon_i(r)$.

References

[1] N. Bloembergen, *Nonlinear Optics*, World Scientific, Singapore, New Jersey, London, Hong Kong, 4 edition, 1996.

[2] P. N. Butcher and D. Cotter, *The Elements of Nonlinear Optics*, Cambridge Univ. Press, New York, 1990.

[3] D. C. Rodenberger, J. R. Heflin, and A. F. Garito, *Nature (London)* **359**, 309 (1992).

[4] G. L. Fischer et al., *Phys. Rev. Lett.* **74**, 1871 (1995).

[5] R. S. Bennink, Y.-K. Yoon, R. W. Boyd, and J. E. Sipe, *Opt. Lett.* **24**, 1416 (1999).

[6] T. Sekikawa, A. Kosuge, T. Kanai, and S. Watanabe, *Nature (London)* **432**, 605 (2004).

[7] J. P. Huang and K. W. Yu, *Appl. Phys. Lett.* **86**, 041905 (2005).

[8] V. M. Shalaev, editor, *Optical Properties of Nanostructured Random Media*, Springer, Berlin, 2002.

[9] V. M. Shalaev, *Phys. Rep.* **272**, 61 (1996).

[10] A. Sarychev and V. Shalaev, *Phys. Rep.* **335**, 275 (2000).

[11] N. N. Lepeshkin, A. Schweinsberg, G. Piredda, R. S. Bennink, and R. W. Boyd, *Phys. Rev. Lett.* **93**, 123902 (2004).

[12] L. Dong, G. Q. Gu, and K. W. Yu, *Phys. Rev. B* **67**, 224205 (2003).

[13] L. Gao, J. P. Huang, and K. W. Yu, *Phys. Rev. B* **69**, 075105 (2004).

[14] J. P. Huang and K. W. Yu, *Appl. Phys. Lett.* **85**, 94 (2004).

[15] J. P. Huang, L. Dong, and K. W. Yu, *Europhys. Lett.* **67**, 854 (2004).

[16] J. P. Huang and K. W. Yu, *Opt. Lett.* **30**, 275 (2005).

[17] A. M. Freyria, E. Chignier, J. Guidollet, and P. Louisot, *Biomaterials* **12**, 111 (1991).

[18] H. Karacali, S. M. Risser, and K. F. Ferris, *Phys. Rev. E* **56**, 4286 (1997).

[19] U. Levy et al., *J. Opt. Soc. Am. A* **22**, 724 (2005).

[20] R. J. Gehr and R. W. Boyd, *Chem. Mater.* **8**, 1807 (1996).

[21] V. M. Shalaev, E. Y. Poliakov, and V. A. Markel, *Phys. Rev. B* **53**, 2437 (1996).

[22] D. R. Smith, J. B. Pendry, and M. C. K. Wiltshire, *Science* **305**, 788 (2004).

[23] J. C. M. Garnett, *Philos. Trans. R. Soc. London Ser. A* **203**, 385 (1904).

[24] J. C. M. Garnett, *Philos. Trans. R. Soc. London Ser. A* **205**, 237 (1906).

[25] D. A. G. Bruggeman, *Ann. Phys. (Leipzig)* **24**, 636 (1935).

[26] D. J. Bergman, *Phys. Rep.* **43**, 377 (1978).

[27] K. P. Yuen and K. W. Yu, *J. Phys.: Condens. Mater* **9**, 4669 (1997).

[28] J. P. Huang and K. W. Yu, *J. Phys.: Condens. Matter* **14**, 1213 (2002).

[29] F. Caruso, *Colloids and colloid assemblies*, Wiley-VCH, Weinheim, 2004.

[30] K. P. Velikov, C. G. Christova, R. P. A. Dullens, and A. V. Blaaderen, *Science* **296**, 106 (2002).

[31] T. Gong and D. W. M. Marr, *Appl. Phys. Lett.* **85**, 3760 (2004).

[32] T. Schilling and D. Frenkel, *Phys. Rev. Lett.* **92**, 085505 (2004).

[33] B. V. R. Tata, P. S. Mohanty, M. C. Valsakumar, and J. Yamanaka, *Phys. Rev. Lett.* **93**, 268303 (2004).

[34] P. Schall, I. Cohen, D. A. Weitz, and F. Spaepen, *Science* **305**, 1944 (2004).

[35] J. H. Holtz and S. A. Asher, *Nature (London)* **389**, 829 (1997).

[36] Y. A. Vlasov, X. Z. Bo, J. C. Strum, and D. J. Norris, *Nature (London)* **414**, 289 (2001).

[37] J. P. Huang and K. W. Yu, *Appl. Phys. Lett.* **87**, 071103 (2005).

[38] C. L. Nehl et al., *Nano Lett.* **4**, 2355 (2004).

[39] D. B. Mitzi, L. L. Kosbar, C. E. Murray, M. Copel, and A. Afzali, *Nature (London)* **428**, 299 (2004).

[40] C. K. Lo and K. W. Yu, *Phys. Rev. E* **64**, 031501 (2001).

[41] L. D. Landau, E. M. Lifshitz, and L. P. Pitaevskii, *Electrodynamics of Continuous Media*, Pergamon Press, New York, 2 edition, 1984.

[42] D. Wang et al., *Small* **1**, 122 (2005).

[43] G. W. Milton, *The Theory of Composites*, Cambridge University Press, Cambridge, England, 2002.

[44] H. Ma, R. Xiao, and P. Sheng, *J. Opt. Soc. Am. B* **15**, 1022 (1998).

[45] A. V. Blaaderen, *MRS Bull.* **29**, 85 (2004).

[46] L. Dong, J. P. Huang, K. W. Yu, and G. Q. Gu, *J. Appl. Phys.* **95**, 621 (2004).

[47] L. Gao, J. P. Huang, and K. W. Yu, *Eur. Phys. J. B* **36**, 475 (2003).

[48] R. J. Elliot, J. A. Krumhansl, and P. L. Leath, *Rev. Mod. Phys.* **46**, 465 (1974).

[49] S. Riikonen, I. Romero, and F. J. G. de Abajo, *Phys. Rev. B* **71**, 235104 (2005).

[50] J. J. Xiao and K. W. Yu, *Appl. Phys. Lett.* , accepted (2005).

[51] J. Lekner, *Physica A* **157**, 826 (1989).

[52] J. Lekner, *Physica A* **176**, 485 (1991).

[53] S. V. Lishchuk, *Molecular Physics* **100**, 3789 (2002).

[54] G. Q. Gu and K. W. Yu, *Phys. Rev. B* **46**, 4502 (1992).

[55] X. C. Zeng, D. J. Bergman, P. M. Hui, and D. Stroud, *Phys. Rev. B* **38**, 10970 (1988).

[56] A. Sharkawy et al., *Opt. Express* **13**, 2814 (2005).

[57] L. Dong, M. Karttunen, and K. W. Yu, *Phys. Rev. E* **72**, 016613 (2005).

[58] D. R. Kammler et al., *J. Appl. Phys.* **90**, 5979 (2001).

[59] S. G. Lu et al., *Appl. Phys. Lett.* **82**, 2877 (2003).

[60] M. Yamanouchi, M. Koizumi, T. Hirai, and I. Shioda, editors, *Proceedings of the First International Symposium on Functionally Graded Materials* , Sendi, Japan, 1990.

[61] H. Grull, A. Schreyer, N. F. Berk, C. F. Majkrzak, and C. C. Han, *Europhys. Lett.* **50**, 107 (2000).

[62] J. D. Jackson, *Classical Electrodynamics*, Wiley, New York, 1975.

[63] G. Q. Gu and K. W. Yu, *J. Appl. Phys.* **94**, 3376 (2003).

[64] K. W. Yu, G. Q. Gu, and J. P. Huang, URL http://www.arxiv.org/pdf/cond-mat/0211532.

[65] J. P. Huang, K. W. Yu, G. Q. Gu, and M. Karttunen, *Phys. Rev. E* **67**, 051405 (2003).

[66] D. Stroud and P. M. Hui, *Phys. Rev. B* **37**, 8719 (1988).

[67] G. S. Agarwal and S. D. Gupta, *Phys. Rev. A* **38**, 5678 (1988).

[68] D. J. Bergman, *Phys. Rev. B* **39**, 4598 (1989).

[69] B. K. P. Scaife, *Principles of Dielectrics*, Calvendon, Oxford, 1989.

[70] D. J. Bergman and D. Stroud, *Solid State Physics: Applied in Research and Applications*, volume 46, page 147, Academic Press, New York, 1992.

[71] K. W. Yu, P. M. Hui, and D. Stroud, *Phys. Rev. B* **47**, 14150 (1993).

[72] R. W. Boyd and J. E. Sipe, *J. Opt. Soc. Am. B* **11**, 297 (1994).

[73] V. M. Shalaev, *Nonlinear Optics of Random Media: Fractal Composites and Metal-Dielectric Films*, Springer, Berlin, 2000.

[74] L. Gao, J. T. K. Wan, K. W. Yu, and Z. Y. Li, *J. Appl. Phys.* **88**, 1893 (2000).

[75] P. M. Hui and D. Stroud, *J. Appl. Phys.* **82**, 4740 (1997).

[76] P. M. Hui, C. Xu, and D. Stroud, *Phys. Rev. B* **69**, 014202 (2004).

[77] P. M. Hui, C. Xu, and D. Stroud, *Phys. Rev. B* **69**, 014203 (2004).

[78] D. Pezzetta et al., *J. Opt. Soc. Am. B* **19**, 2102 (2002).

[79] G. Purvinis et al., *Opt. Lett.* **29**, 1108 (2004).

[80] R. W. Boyd, *Nonlinear Optics*, Academic Press, New York, 1992.

[81] H. Spcker, M. Portun, and U. Woggon, *Opt. Lett.* **23**, 427 (1998).

[82] S. Yu, J. I. Lee, and A. K. Viswanath, *J. Appl. Phys.* **86**, 3159 (1999).

[83] P. Fu, Q. Jiang, X. Mi, and Z. Yu, *Phys. Rev. Lett.* **88**, 113902 (2002).

[84] A. E. Neeves and M. H. Birnboim, *J. Opt. Soc. Am. B* **6**, 787 (1989).

[85] B. Pettinger, X. Bao, I. C. Wilcock, M. Muhler, and G. Ertl, *Phys. Rev. Lett.* **72**, 1561 (1994).

[86] H.-P. Chiang, P. T. Leung, and W. S. Tse, *J. Phys. Chem. B* **104**, 2348 (2000).

[87] P. M. Hui, P. C. Cheung, and D. Stroud, *J. Appl. Phys.* **84**, 3451 (1998).

[88] J. P. Huang and K. W. Yu, *J. Appl. Phys.* **97**, 013102 (2005).

[89] J. P. Huang and K. W. Yu, *J. Appl. Phys.* , to be accepted (2006).

[90] K. W. Yu, P. M. Hui, and H. C. Lee, *Phys. Lett. A* **210**, 115 (1996).

[91] T. B. Jones, *Electromechanics of Particles*, Cambridge University Press, Cambridge, England, 1995.

[92] P. Hui, X. Zhang, and D. Stroud, *J. Mater. Sci.* **34**, 5497 (1999).

[93] E. Wei, Y. Poon, and F. Shin, *Phys. Lett. A* **336**, 264 (2005).

[94] Z. F. Sang and Z. Y. Li, *Phys. Lett. A* **334**, 422 (2005).

[95] J. W. Haus, R. Inguva, and C. M. Bowden, *Phys. Rev. A* **40**, 5729 (1989).

[96] J. E. Sipe and R. W. Boyd, *Phys. Rev. A* **46**, 1614 (1992).

[97] K. P. Yuen and K. W. Yu, *J. Opt. Soc. Am. B* **14**, 1387 (1997).

[98] M. Tlidi, M. F. Hilali, and P. Mandel, *Europhys. Lett.* **55**, 26 (2001).

[99] P. Mulvaney, *MRS Bull.* **26**, 1009 (2001).

[100] S. Roorda et al., *Adv. Mater.* **16**, 235 (2004).

[101] C. Sönnichsen et al., *Phys. Rev. Lett.* **88**, 077402 (2002).

[102] K. P. Yuen, M. F. Law, and K. W. Yu, *Phys. Rev. E* **56**, R1322 (1997).

[103] B. M. I. V. D. Zande, L. Pages, R. A. M. Hikmet, , and A. V. Blaaderen, *J. Phys. Chem. B* **103**, 5761 (1999).

[104] E. Snoeks et al., *Adv. Mater.* **12**, 1511 (2000).

[105] A. Benyagoub et al., *Nucl. Instr. Methods Phys. Res. B* **64**, 684 (1992).

[106] B. Xu, J. P. Huang, and K. W. Yu, *Phys. Lett. A* , submitted (2005).

[107] R. S. Bennink, Y.-K. Yoon, R. W. Boyd, and J. E. Sipe, *Opt. Lett.* **24**, 1416 (1999).

[108] W. T. Wang et al., *Appl. Phys. Lett.* **83**, 1983 (2003).

[109] G. Q. Gu, P. M. Hui, and K. W. Yu, *Physica B* **279**, 62 (2000).

[110] E. B. Wei, J. B. Song, and G. Q. Gu, *J. Appl. Phys.* **95**, 1377 (2004).

[111] E. B. Wei, Z. D. Yang, and G. Q. Gu, *J. Phys. D: Appl. Phys.* **37**, 107 (2004).

[112] J. P. Huang, *J. Phys. Chem. B* **109**, 4824 (2005).

[113] L. Gao, *Phys. Rev. E* **71**, 067601 (2005).

[114] D. Stroud and V. E. Wood, *J. Opt. Soc. Am. B* **6**, 778 (1989).

[115] C. F. Bohren and D. R. Huffman, *Absorption and Scattering of Light by Small Particles*, John Wiley & Sons, New York, 1983.

[116] Z. Hashin and S. Shtrikman, *J. Appl. Phys.* **33**, 3125 (1962).

[117] J. P. Huang and K. W. Yu, *J. Opt. Soc. Am. B* **22**, 1640 (2005).

[118] K. W. Yu, *Solid State Commun.* **105**, 689 (1998).

[119] I. Tanahashi, Y. Manabe, T. Tohda, S. Sasaki, and A. Nakamura, *J. Appl. Phys.* **79**, 1244 (1996).

[120] J. P. Huang, P. M. Hui, and K. W. Yu, *Phys. Lett. A* **342**, 484 (2005).

[121] L. Gao, J. P. Huang, and K. W. Yu, *Phys. Rev. E* **67**, 021910 (2003).

[122] L. Gao, J. P. Huang, and K. W. Yu, *Phys. Rev. B* **69**, 075105 (2004).

[123] L. Gao and Z. Y. Li, *J. Appl. Phys.* **91**, 2045 (2002).

[124] J. P. Huang, L. Gao, K. W. Yu, and G. Q. Gu, *Phys. Rev. E* **69**, 036605 (2004).

[125] L. Gao, K. W. Yu, Z. Y. Li, and B. Hu, *Phys. Rev. E* **64**, 036615 (2001).

[126] K. W. Yu and G. Q. Gu, *Phys. Lett. A* **193**, 311 (1994).

[127] O. Levy, D. J. Bergman, and D. Stroud, *Phys. Rev. E* **52**, 3184 (1995).

[128] L. Gao and K. W. Yu, *Phys. Rev. B* **72**, 075111 (2005).

[129] Y. Hamanaka, K. Fukuta, A. Nakamura, L. Liz-Marzan, and P. Mulvaney, *Appl. Phys. Lett.* **84**, 4938 (2004).

[130] L. Gao and K. W. Yu, *Phys. Rev. E* **71**, 017601 (2005).

[131] M. Avellaneda, A. Cherkaev, K. Lurie, and G. Milton, *J. Appl. Phys.* **63**, 4989 (1988).

[132] J. H. Erdmann, S. Zumer, and J. Doane, *Phys. Rev. Lett.* **64**, 1907 (1990).

[133] V. L. Sukhorukov, G. Meedt, M. Kürschner, and U. Zimmermann, *J. Electrost.* **50**, 191 (2001).

[134] D. Stroud, *Phys. Rev. B* **54**, 3295 (1996).

[135] J. Roth and M. J. Dignam, *J. Opt. Soc. Am.* **63**, 308 (1973).

[136] S. Barabash and D. Stroud, *J. Phys.: Condens. Matter* **11**, 10323 (1999).

[137] L. Gao, J. P. Huang, and K. W. Yu, *Eur. Phys. J. B* **36**, 475 (2003).

[138] L. Gao, J. T. K. Wan, K. W. Yu, and Z. Y. Li, *J. Phys.: Condens. Mater* **12**, 6825 (2000).

[139] O. Levy and D. Stroud, *Phys. Rev. B* **56**, 8035 (1997).

[140] G. W. Milton, *Appl. Phys. A* **26**, 1207 (1981).

[141] G. W. Milton, *J. Appl. Phys.* **52**, 5286 (1980).

[142] M. P. Hobson and J. E. Baldwin, *Appl. Opt.* **43**, 2651 (2004).

[143] S. Martin, J. Rivory, and M. Schoenauer, *Appl. Opt.* **34**, 2247 (1995).

[144] P. G. Verly, *Appl. Opt.* **37**, 7327 (1998).

[145] Y. Gu and Q. H. Gong, *Phys. Rev. B* **67**, 014209 (2003).

[146] O. Levy and D. J. Bergman, *Physica A* **207**, 157 (1994).

[147] H. Ma, B. Zhang, W. Y. Tam, and P. Sheng, *Phys. Rev. B* **61**, 962 (2000).

[148] X. Zhang and D. Stroud, *Phys. Rev. B* **49**, 944 (1994).

[149] R. E. Rosensweig, *Ferrohydrodynamics*, Cambridge University Press, Cambridge, England, 1985.

[150] B. M. Berkovsky, V. F. Medvedev, and M. S. Krakov, *Magnetic Fluids, Engineering Applications*, Oxford University Press, Oxford, England, 1993.

[151] S. Odenbach, *Magn. Elect. Separation* **9**, 1 (1998).

[152] R. Hergt et al., *IEEE Trans. Magn.* **34**, 3745 (1998).

[153] C. Alexiou et al., *Magnetohydrodynamics* **37**, 3 (2001).

[154] S. Odenbach, *Magnetoviscous Effects in Ferrofluids*, Springer, Berlin, 2002.

[155] H. Bönnemann et al., *Inorganica Chimica Acta* **350**, 617 (2003).

[156] J. E. Martin, R. A. Anderson, and C. P. Tigges, *J. Chem. Phys* **108**, 3765 (1998).

[157] J. E. Martin, R. A. Anderson, and C. P. Tigges, *J. Chem. Phys* **108**, 7887 (1998).

[158] J. P. Huang, *Phys. Rev. E* **70**, 041403 (2004).

[159] M. Rasa, *J. Magn. Magn. Mater.* **201**, 170 (1999).

[160] J. I. Dadap, J. Shan, K. B. Eisenthal, and T. F. Heinz, *Phys. Rev. Lett.* **83**, 4045 (1999).

[161] N. Yang, W. E. Angerer, and A. G. Yodh, *Phys. Rev. Lett.* **87**, 103902 (2001).

[162] P. Xu et al., *Phys. Rev. Lett.* **93**, 133904 (2004).

[163] R. Bernal and J. A. Maytorena, *Phys. Rev. B* **70**, 125420 (2004).

[164] V. F. Puntes, P. Gorostiza, D. M. Aruguete, N. G. Bastus, and A. P. Alivisatos, *Nat. Mater.* **3**, 263 (2004).

[165] Y. R. Shen, *The Principles of Nonlinear Optics*, J. Wiley and Sons, New York, 1984.

[166] M. I. Stockman, D. J. Bergman, C. Anceau, S. Brasselet, and J. Zyss, *Phys. Rev. Lett.* **92**, 057402 (2004).

[167] P. G. Kik, S. A. Maier, and H. A. Atwater, *Phys. Rev. B* **69**, 045418 (2004).

[168] T. Du and W. Luo, *Appl. Phys. Lett.* **72**, 272 (1998).

[169] M. Whittle and W. A. Bullough, *Nature* **358**, 373 (1992).

[170] T. C. Halsey, *Science* **258**, 761 (1992).

[171] W. J. Wen, X. X. Huang, S. H. Yang, K. Q. Lu, and P. Sheng, *Nature Materials* **2**, 727 (2003).

[172] W. M. Winslow, *J. Appl. Phys.* **20**, 1137 (1949).

[173] R. Tao and J. M. Sun, *Phys. Rev. Lett.* **67**, 398 (1991).

[174] R. Tao, editor, *Electrorheological Fluids*, World Scientific, Singapore, 1992.

[175] R. Tao and G. D. Roy, editors, *Electrorheological Fluids*, World Scientific, Singapore, 1994.

[176] W. A. Bullough, editor, *Electro-Rheological Fluids, Magneto-Rheological Suspensions and Associated Technology*, World Scientific, Singapore, 1996.

[177] G. Bossis, editor, *Electrorheological Fluids and Magnetorheological Suspensions*, World Scientific, Singapore, 2001.

[178] B. Hoffmann and W. Köhler, *J. Magn. Magn. Mater.* **262**, 289 (2003).

[179] C. Park and R. E. Robertson, *Mat. Sci. Eng. A - Struct.* **257**, 295 (1998).

[180] A. Kawai, K. Uchida, and F. Ikazaki, *Electrorheological Fluids and Magetorheological Suspensions*, pages 626–632, World Scientific, Singapore.

[181] J. P. Huang, J. T. K. Wan, C. K. Lo, and K. W. Yu, *Phys. Rev. E* **64**, 061505(R) (2001).

[182] J. Messier, F. Kajzar, P. Prasad, and D. Ulrich, editors, *Nonlinear optical effects in organic polymers*, Springer, Berlin, 1989.

[183] I. D. L. Albert, T. J. Marks, and M. A. Ratner, *J. Am. Chem. Soc.* **120**, 11174 (1998).

[184] M. Ahlheim et al., *Science* **271**, 335 (1996).

[185] E. Rosencher et al., *Science* **271**, 168 (1996).

[186] C. Dhenaut et al., *Nature (London)* **374**, 339 (1995).

[187] I. D. L. Albert, T. J. Marks, and M. A. Ratner, *J. Am. Chem. Soc.* **119**, 6575 (1997).

[188] I. D. L. Albert, T. J. Marks, and M. A. Ratner, *Chem. Mater.* **10**, 753 (1998).

[189] H. S. Nalwa and S. Miyata, *Nonlinear Optics of Organic Molecules and Polymers*, CRC Press, Boca Raton, New York, London, Tokyo, 1997.

[190] M. G. Kuzyk and C. W. Dirk, *Characterization Techniques and Tabulations for Organic Nonlinear Optical Materials*, Marcel Dekker, Inc., New York, Basel, Hong Kong, 1998.

[191] M. Y. Antipin et al., *J. Mater. Chem.* **11**, 351 (2001).

[192] P. Günter, C. Bosshard, K. Sutter, and H. Arend, *Appl. Phys. Lett.* **50**, 486 (1987).

[193] S. Tomary et al., *Appl. Phys. Lett.* **58**, 2583 (1991).

[194] Y. L. Fur, M. Bagieu-Beucher, R. Masse, J.-F. Nicoud, and J.-P. Levy, *Chem. Mater.* **8**, 68 (1996).

[195] M. Eich et al., *J. Opt. Soc. Am. B* **6**, 1590 (1989).

[196] J. Pecaut, J. P. Levy, and R. Masse, *J. Mater. Chem.* **3**, 999 (1993).

[197] A. Yokoo, S. Tomaru, I. Yokohama, H. Itoh, and T. Kaino, *J. Cryst. Growth* **156**, 279 (1995).

[198] S. H. Tolbert, A. Firouzi, G. D. Stucky, and B. F. Chmelka, *Science* **278**, 264 (1997).

[199] T. Verbiest et al., *Science* **282**, 913 (1998).

[200] D. Jungbauer et al., *J. Appl. Phys.* **69**, 8011 (1991).

[201] K. S. Suslick, C. T. Chen, G. R. Meredith, and L. T. Cheng, *J. Am. Chem. Soc.* **114**, 6928 (1992).

[202] M. Ohkita, T. Suzuki, K. Nakatani, and T. Tsuji, *Chem. Commun.* , **1454** (2001).

[203] O. R. Evans and W. Lin, *Acc. Chem. Res.* **35**, 511 (2002).

[204] A. L. Roy et al., *Comptes Rendus Chimie* **8**, 1256 (2005).

[205] H. Koshima, H. Miyamoto, I. Yagi, and K. Uosaki, *Molecular Crystals and Liquid Crystals* **420**, 79 (2004).

[206] V. V. Nesterov et al., *J. Phys. Chem. B* **108**, 8531 (2004).

[207] M. D. Aggarwal, J. Stephens, A. K. Batra, and R. B. Lal, *Journal of Optoelectronics and Advanced Materials* **5**, 555 (2003).

[208] K. V. Katti et al., *Chem. Mater.* **14**, 2436 (2002).

[209] O. R. Evans and W. B. Lin, *Chem. Mater.* **13**, 2705 (2001).

[210] O. R. Evans and W. B. Lin, *Chem. Mater.* **13**, 3009 (2001).

[211] S. Kosaka, Y. Benino, T. Fujiwara, V. Dimitrov, and T. Komatsu, *J. Solid State Chem.* **178**, 2067 (2005).

[212] H. Kishida et al., *Nature (London)* **405**, 929 (2000).

[213] H. Kagawa et al., *Chem. Mater.* **8**, 2622 (1996).

[214] A. L. Stepanov et al., *Tech. Phys. Lett.* **30**, 846 (2004).

[215] Y. Mori, I. Kuroda, S. Nakajima, T. Sasaki, and S. Nakai, *Appl. Phys. Lett.* **67**, 1818 (1995).

[216] W. R. Bosenberg, L. K. Cheng, and J. D. Bierlein, *Appl. Phys. Lett.* **65**, 2765 (1994).

[217] C. C. Evans, M. Bagieu-Beucher, R. Masse, and J. F. Nicoud, *Chem. Mater.* **10**, 847 (1998).

[218] P. A. A. Mary and S. Dhanuskodi, *Crystal Res. Tech.* **36**, 1231 (2001).

[219] B. J. Scott, G. Wirnsberger, and G. D. Stucky, *Chem. Mater.* **13**, 3140 (2001).

[220] A. P. Alivisatos, *J. Phys. Chem.* **100**, 13226 (1996).

[221] M. Ogawa, T. Nakamura, J. Mori, and K. Kuroda, *J. Phys. Chem. B* **104**, 8554 (2000).

[222] J. P. Huang, *Phys. Rev. E* **70**, 042501 (2004).

[223] E. C. Stoner, *Philos. Mag.* **36**, 308 (1945).

[224] J. A. Osborn, *Phys. Rev.* **67**, 351 (1945).

[225] W. T. Doyle and I. S. Jacobs, *J. Appl. Phys.* **71**, 3926 (1992).

[226] Z. Hashin, *J. Appl. Mech.* **29**, 143 (1962).

[227] K. W. Yu, J. T. K. Wan, M. F. Law, and K. K. Leung, *Int. J. Mod. Phys. C* **9**, 1447 (1998).

[228] K. Schulgasser, *J. Appl. Phys.* **54**, 1380 (1983).

[229] K. W. Yu and G. Q. Gu, *Phys. Lett. A* **345**, 448 (2005).

[230] R. Rojas, F. Claro, and R. Fuchs, 37, 6799 (1988).

[231] L. Dong, J. P. Huang, K. W. Yu, and G. Q. Gu, *Eur. Phys. J. B*, accepted (2005).

[232] A. A. Lucas, L. Henrard, and P. Lambin, *Phys. Rev. B* **49**, 2888 (1994).

Index

O

Q

R

P

S